Software and the Agile Manufacturer

Software and the Agile Manufacturer

Computer Systems and World Class Manufacturing

Brian H. Maskell

Productivity Press

PORTLAND, OREGON

Productivity Press
P.O. Box 13390
Portland, Oregon 97313
United States of America
Telephone: 1-800-394-6868
Telefax: 1-800-394-6286

Cover design by Teutschel Design Services
Printed and bound by BookCrafters USA
Printed in the United States of America

Library of Congress Cataloging-in-Publication Data

Maskell, Brian
    Software and the agile manufacturer: computer systems and world class manufac-
turing / by Brian H. Maskell.
        p.   cm.
    Includes bibliographical references and index.
    ISBN 1-56327-046-3
    1. Product planning--Data processing.   2. Manufacturing resource planning--Data
processing. 3. Computer integrated manufacturing systems. 4. Just-in-time systems.
I. Title.
TS176.M3645   1993   93-33282
658.5'036—dc20   CIP

97 96 95 94 9 8 7 6 5 4 3 2 1

*This book is dedicated to my sons
Sam and Caleb with much love.*

Contents

Publisher's Message

At last someone has written an extremely practical book to help managers of manufacturing companies cope with the confusion of ever-changing computer technology and software. *Software and the Agile Manufacturer: Computer Systems and World Class Manufacturing* addresses this issue, one that author Brian H. Maskell understands well, having been in on ground-floor developments for years. He knows from experience that manufacturing and distribution organizations moving from traditional manufacturing and distribution techniques to more agile, world-class approaches find this transition (which is difficult enough) usually hindered by the computer systems involved. He provides us with guidelines for developing systems, or modifying existing ones, that aid in the transition. Although he deals with some new technologies helpful to Western manufacturers, his book is not about technology per se. It is about how to manage company-wide change utilizing one's current software systems.

Examining American success stories like Xerox, General Motors, Harley-Davidson, Hewlett-Packard, Motorola, Digital Equipment, Federal Express, and Caterpillar as well as companies in other countries, the author helps us makes sense of an intimidating situation. In short, he shows us that software can be friendly.

Software and the Agile Manufacturer is about how to use software in the transition process. It begins by exploring various aspects of agile manufacturing and then examines the relevance of software systems in each area. A practical analysis of software approaches follows. The author recognizes that it is often impossible to throw away systems developed over many years and replace them with something new. Instead he seeks to provide some guidance to manufacturing managers and MIS people engaged in the thankless task of having to make do with inadequate systems. Let me summarize the chapters:

Chapter 1: While MRP II and agile manufacturing objectives are entirely congruent, MRP II contains a rigid operations-research approach to solving production and inventory problems that needs changing for it to be useful in an agile environment. Although software-support needs vary considerably among companies, some underlying principles of software design include the elements of integration, simplicity, flexibility, openness, and accessibility.

Chapter 2: New methods of manufacturing and distribution (known collectively as "agile manufacturing") seek to bring about the radical changes required for a company to be competitive on the world market. This usually encompasses a new approach to quality, JIT production techniques, vendor relationships, and managing people, as well as the development of more flexible production methods and more rapid introduction of new products.

Chapter 3: Here follows a discussion of the issues at the heart of agile production planning processes: cellular manufacturing, rate-based scheduling, and capacity planning.

Computer systems must contain features that assist in the transition to these new approaches.

Chapter 4: Executing production plans on the shop floor differs fundamentally when agile techniques are used. The purpose is to minimize inventory, move materials on a JIT basis, and eliminate unnecessary transactions. Software must support backflushing, inventory pull, tracking lot and serial numbers, and kanban cards.

Chapter 5: Computer systems must be customer oriented, respond to customer needs, and provide flexibility. The goal is the just-in-time delivery of make-to-order products with no delays and virtually no administative procedures. Systems are integrated, simple, yet focused to the needs of the customer and the company. While order-entry processes vary by product and according to the marketing strategy, the complex, multi-step allocation, picking, staging, packing, and dispatch processes common in traditional companies are simplified and largely eliminated.

Chapter 6: Three aspects of computer systems are increasingly prominent within agile manufacturers: order-entry configurator, electronic data interchange (EDI), and distribution resource planning. Previously considered advanced, these methods are being accepted as a standard part of the new procedures.

Chapter 7: Software support must substantially simplify the traditional approach to procurement, which is largely based on old ideas of adversarial company-vendor relationships. Here the author examines blanket or contract purchase orders and the use of supplier kanbans, faxed orders, or electronic signals to simplify the call-off process. Invoicing can be replaced by the automatic creation of invoice records from backflushing, advanced shipping notices (ASNs), or receiving transactions.

Chapter 8: Performance measures in agile manufacturing incorporate seven common characteristics: they directly relate to the manufacturing strategy, they primarily use nonfinancial measures, they vary among locations, they change over time, they are simple and easy to use, they provide fast feedback to operators and managers, and they foster improvement instead of simply monitoring. Although manual reporting is common, the use of computer systems to provide performance reports is still important. Shop-floor terminals, local PCs or workstations for analyzing or presenting the information, and fourth-generation report writers help achieve this goal.

Chapter 9: Here we look at traditional accounting systems, which many agile manufacturers find difficult to remove or modify because they are so much a part of the company's infrastructure. The best approach to management-accounting systems is to largely eliminate them from the day-to-day operation of the production and distribution plant and to run the plant using nonfinancial reporting and control.

Chapter 10 looks at the agile manufacturing's objectives of product design — to quickly develop and bring to market innovative new products that exceed customer expectations and desires, have high quality and low cost, and can be manufactured easily. These objectives are achieved through the use of cross-functional design teams employing the techniques of concurrent engineering, computer-aided design (CAD), and computer-aided process planning systems (CAPP), quality function deployment (QFD), design-for-manufacturability analyses, and target costing and value engineering.

Chapter 11 examines the emerging trends of graphical user interface (GUI), relational data bases, networks and client/server arrangements, software useable on many different computers, and "open" computer systems. Computer-integrated manufacturing (CIM) and the use of robots, numerically controlled (NC) machines, and automated warehousing and material-handling systems are not cures in themselves — but their appropriate use can be a powerful competitive edge. Data collection, often using barcodes, is an effective way to improve accuracy of transaction entries and to reduce the burden of data entry.

It is always a great pleasure for me to introduce a new Productivity Press book to the public. To Brian Maskell goes my heartfelt gratitude for permitting us to publish more of his exceptional work. In turn, I also thank the many fine people who contributed to this project, among them Ron Bridenthal, freelance editor (Carol Stream, Illinois); Karen Jones, managing editor; Jennifer Albert, editorial assistant; Bill Stanton, production manager; and Cheryl Rosen, senior editor and project manager.

Norman Bodek
CEO, Productivity Press

Preface

*S*oftware and the Agile Manufacturer is about the management of manufacturing and distribution organizations in the process of making the difficult transition from traditional manufacturing and distribution techniques to an agile manufacturing approach. This transition is complex and difficult, and the computer systems involved are often a hindrance rather than a help. This book provides some guidelines for the development of systems, or modification of existing ones, so they become truly helpful in the quest for world-class excellence.

The term "agile manufacturing" first reached a wide audience when Lehigh University published its influential report, *21st Century Manufacturing Enterprise Strategy*. This report outlined the radical changes required for American industry to be competitive in the 1990s and into the next century. The introduction of agile manufacturing methods began in the late 1980s and created a quiet revolution that is now affecting the whole of Western industry. Much of this change was initiated under the onslaught of competition from Japanese and other Pacific Rim companies. Competition of this sort created a terrible shock wave in the West, particularly for U.S. companies. Since World War II, America had been the industrial giant, the workshop of the Western world, the leader of the "free" world. During the 1970s and 1980s entire domestic industries were wiped out, and new technologies — often developed by U.S. research — were introduced and exploited by the industrial goliaths from Asia.

I remember clearly the horror that spread through the Xerox Corporation in 1981 when the first set of benchmarks was disseminated. The figures showed that Japanese companies could *sell* an equivalent copier for less than Xerox could *make* it. And this was at a time when everything from cars to cameras to consumer electronics was on its way East. Xerox, the innovator of plain-paper copiers, having enjoyed many years of 15 to 25 percent growth, was on its way out. And we knew it.

Companies like Xerox, Hewlett-Packard, Motorola, Caterpillar, Harley-Davidson, and countless others with lesser-known names have revolutionized their organizations in recent years. They have introduced unheard-of quality standards, innovative just-in-time (JIT) production processes, genuine responsiveness to customer needs, and product-design approaches that "get it right the first time." Western industry was caught off guard in the 1970s and 1980s, but enormous progress has been made since then.

The computer-software industry came of age in the same 1970s and early 1980s. All kinds of companies, government agencies, and service organizations began to use software systems to handle accounts, plan operations, and analyze markets. Production software moved from stock control to material requirements planning (MRP) to manufacturing resource planning (MRP II). MRP II was going to be the tool that would (in the words of pundit Oliver W. Wight) "unlock America's productivity potential." Hundreds of companies developed systems based around the ideas of closed-loop MRP II; many software companies were highly successful at developing and marketing packaged systems of one sort or another. Organizations like the American Production and Inventory Control Society (APICS) and similar bodies in other Western countries flourished with the demand for education and research into approaches to solving the problems of production, inventory, and productivity.

The momentum created by the marketing of an MRP II approach was such that the manufacturing and distribution software industry was, like its customers, caught off guard by the movement toward JIT and agile manufacturing. There was a flurry of dismay and excitement in the mid-1980s when the MRP II approach was first challenged by just-in-timers. Conference speakers (myself included) fell over themselves to demonstrate that MRP II and JIT were fully compatible, or not compatible, or quasi-compatible in a Western context — taking account of the fundamentally different cultural infrastructure of Western manufacturing, and so on and so on.

Consequently, the theme of this book rests on four facts I have found to be true in practice:

1. The majority of Western manufacturers and distributors are working to become "world class." They are going through the painful process of transition from traditional to agile manufacturing methods. There is a diversity of approach: some emphasize quality, some customer service; others, innovation, employee involvement, time-to-market, and so forth. But all are striving for excellence.

2. Developing software to meet the needs of a "fully fledged," mature agile manufacturer is relatively straightforward. However, providing software that supports a company in transition is a more subtle and complex task. The software must be able to support a gradual, smooth, low-risk transition.

3. The introduction of agile manufacturing and distribution methods has very little to do with the software being used. Changes required in an organization striving for world-class status have more to do with leadership and management than with the software systems being employed.
4. Companies successful with agile manufacturing in its many aspects have perfected and simplified their processes to the point where they do not need complex computer systems to run their operations. The ideas of simplicity are reflected in their production systems.

This book is about how software can be used to assist a company making that transition from traditional methods to world-class techniques. We begin by exploring various aspects of agile manufacturing techniques and by examining the relevance of software systems in each of these areas. A very practical analysis of the software approaches that help with the journey to world-class status follows. I recognize that it is not often possible to throw away systems that have been developed over many years and replace them with something new and better. The book seeks to provide some guidance to manufacturing managers and MIS people engaged in the thankless task of having to make do with inadequate systems.

Although *Software and the Agile Manufacturer* deals with some of the new technologies that can be helpful to Western manufacturers, it is not about technology. It is a book about managing the process of transition from traditional to agile manufacturing with an emphasis on how software systems can help this process.

These are challenging times for manufacturers and the people charged with the task of providing computer systems to meet these challenges. With this book I hope to shed some light on the path. Please feel free to get in touch with me if you want to discuss these issues further. I can be reached by telephone at (609) 772-1244 and by fax machine at (609) 427-9560.

Brian H. Maskell
Voorhees, New Jersey
September 1993

Acknowledgments

Writing a book is very much a collaborative effort. I have, of course, spent hundreds of hours in front of the word processor searching for the right thought and the *bon mot*, but I cannot claim the contents of this book to be my own work. It is the result of working for many years with very talented and capable people. One person I would like to pay special tribute to is my friend, colleague, and mentor, Dr. David A. Lilly, who sadly for me passed away this year. Dave worked in manufacturing and distribution software for most of his career with the Xerox Corporation and the Unitronix Corporation. Working with him taught me a great deal about the software industry, how to develop software, how to manage the development team, and how to serve the customer with integrity.

I would also like to acknowledge the people in the many companies I have worked with, both in the United States and in Europe, who have implemented the ideas and concepts discussed in this book. It is through working with these people that these ideas have developed and become tangible.

A big thank you goes to Cheryl Rosen, my editor at Productivity Press, for her help and encouragement; freelance editor Ron Bridenthal, who did sterling work on the manuscript; and all the people at Productivity who have worked to make this book a reality.

And I must, as always, acknowledge my wife Barbara, who constantly supports and encourages me in my endeavors and carries the load when I am away — physically and mentally. Without Barbara's help I could achieve nothing.

CHAPTER ONE

Why MRP II Has Not Created Agile Manufacturing

Where We Go from Here

Computer systems used by the majority of Western manufacturers and distributors are based on the ideas and concepts of a traditional approach to production planning and inventory control. These systems hamper the introduction of agile manufacturing methods and lack the flexibility to allow for the painless introduction of more agile approaches.

The Early Days of Computing

Computers were first used by manufacturing and distribution operations in the 1960s. While by today's standards these systems were rudimentary, at the time they were new and radical. The machines themselves had very little data storage capacity, were slow, and could not handle complex manipulation of either numerical or textual data. As a result, their use in production plants was limited, usually restricted to handling some accounting functions and related stock-control procedures.

In the 1960s manufacturing companies commonly had stockroom inventory transactions recorded on punched cards or paper tape. The resulting stock figures were printed on "tabs" each week. One of the first jobs the author had was as a stockroom supervisor in an electronics manufacturing company that produced marine radar and guidance equipment. One major weekly task was the futile exercise of trying to match manual bin records with the "stock tab." These stock-control reports were used by the purchasing department to help determine the quantities of components to be ordered and by the cost accountants who were interested in inventory valuation.

The more sophisticated of such early stock-control systems incorporated some method of recommending replenishment quantities for raw materials and finished products. These systems were usually based on a reorder quantity (ROQ) and a reorder level (ROL). The ROL for the part was the safety stock the company wished to hold. The ROQ was calculated using demand history and some kind of lot size or economic-order-quantity calculation. Clever forecasting methods, like "single exponential smoothing," calculated a current forecast without taking up precious disk space to store huge histories of demand. The system would then report to the purchasing department items that had fallen below their reorder level and would recommend placing an order for the reorder quantity.

These systems were useful but not very effective. In general, people using the systems could not rely on the accuracy of figures presented in the reports and would frequently resort to manual bin cards or a physical stock count before placing purchase orders. Because they were batch systems, to have a chance of being accurate the transactions were entered according to strict cut-offs for stock levels and inventory valuations. These disciplines 0were difficult to maintain, and the real method of production planning continued to be the hurly-burly of shortage lists, expedite reports, rush jobs, month-end shipping, and manual intervention. The heroes in the plant were the "fire fighters" who got the "hot" orders shipped out in the nick of time, irrespective of the system.

Material Requirements Planning

Material requirements planning (MRP) was introduced in the late 1960s and early 1970s. The pioneers of MRP, Joseph Orlicky and Oliver W. Wight, were looking for better methods of planning and ordering material and components.[1,2] They wanted to order the material based on future planned production rather than on the demand history for a part. Demand history only has value when the future production plan is similar to that of the past. Large amounts of safety stock are required to maintain component availability when production plans and demand vary to any extent. The ROQ and ROL approaches to component replenishment require high inventory levels, resulting in both high costs and the confusion associated with heavily stocked warehouses.

The idea behind MRP is simple. If you know what the production plan will be for final assemblies, if you know the bill of material (BOM) required to make each product, and if you know what component inventories you have, you can calculate what you need and when you need it. The process of netting inventory is straightforward; component quantity in a BOM is multiplied by the quantity to be made, thus giving the gross requirement. Current available inventory is then subtracted from the gross requirement to obtain the net requirement, which is the quantity to be purchased to complete the production job. This calculation becomes more complex when the same components are used on many products and when bills of material are deep and complex. However, the principle of the MRP calculation is the same. While these calculations were performed manually in earlier times, computers made it possible for calculations to be done quickly and to reflect the most recent information.

Early MRP systems calculated in monthly "time buckets." In other words, they calculated the requirement for an entire month of production, and the components were purchased for that month. As the MRP approach became more widely used, the process was reduced to weekly buckets and then to daily buckets. MRP becomes a more powerful tool when daily buckets are used because it ceases to be merely a materials planning system — it can also be used to schedule shop-floor production.

MRP reports could show when the "due date" of a released order was not in line with the customer's "required date." These production orders could be sequenced and prioritized, and these priorities could be adjusted as the situation changed day by day. This procedure represented a real breakthrough in manufacturing production planning and control. MRP provided production controllers with an automated mechanism for priority scheduling of the shop floor, giving them — for the first time — genuine control of shop-floor production that was matched to customer requirements.

Capacity Requirements Planning

Scheduling the shop floor and purchasing components and materials is only one side of the production-planning problem. The other side is the availability of production capacity. After MRP systems had been used by a number of companies and had gained acceptance within the production and inventory-control community, the innovators turned their attention to capacity planning.

Capacity requirements planning (CRP) is similar to MRP except that this system deals with production time instead of components and materials. The work load required to manufacture an item is calculated by multiplying the quantity to be made by the "bill of labor" for that product. The bill of labor is more commonly known as a production routing. The production routing contains details of how the product is made, which work centers are required, how many setups are needed, the time to process one item, the move times as the product journeys through the production process, and the amount of time spent queuing on the shop floor.

The bill of material and production routing together constitute the production-engineering definition of the product because they describe what the product contains and how it is put together. The success of capacity planning requires high levels of accuracy in the definition of production routings, just as MRP requires accurate BOMs. The capacity required to complete daily or weekly production, once calculated, can be compared to the available capacity within each work center. The CRP system will highlight work centers that are overloaded on any particular day.

Production planners, armed with the MRP schedule information and the CRP information, can then use their judgment to replan and reschedule the production plant to ensure the most effective use of materials and capacity. These decisions take account of each order's priority, the customer's needs and terms of business, the availability of additional capacity through overtime or changing work crews, and other aspects.

In recent years this approach has been refined by combining the MRP and CRP calculations so that detailed scheduling is performed within capacity planning, taking account of capacity constraints. This approach, known as finite capacity planning, attempts to match available resources with prioritized demand upon those resources.

Finite scheduling reschedules production orders according to available capacity and reschedules component and material requirements according to the actual daily requirement of the finite schedule.

Finite scheduling has had limited success because available capacity is difficult to define accurately. People often can be quite flexible when there are unusual demands upon them and it is difficult for capacity planning to accurately account for such issues as overtime, reorganized work crews, additional outside processing, and alternate production routings. Finite scheduling has been more successful within process manufacturers where these constraints can be more clearly defined and are less flexible.

Some innovative approaches to the solution of this problem have emerged. One that has been quite successful is optimized production technique (OPT). Using the concepts of constraints theory, OPT determines which work centers are the bottlenecks in the process. These work centers are then finite-scheduled to ensure maximum flow of production through the bottleneck. In turn, production flow is optimized plantwide. Finite scheduling of the entire plant is not necessary if the bottleneck work centers are handled correctly. While this approach has been highly successful in some kinds of production plants, it is not a panacea.

Manufacturing Resource Planning

The introduction in the mid-1970s of manufacturing resource planning (MRP II) was heralded as the approach that would enable Western manufacturers to attain world-class status. Oliver Wight called MRP II the method for "unlocking America's productivity potential."[3] MRP II could be performed using a computer, and the computers available at that time had the power, speed, and storage capacity to handle the volumes of data and calculations required by MRP II.

MRP II takes the approach that a company is not a series of independent activities. Rather it is an integrated set of activities that all impact upon one another. To be effective, a computer system must also be integrated and take account of changes occurring within the company on a daily basis. MRP II takes the ideas of MRP and CRP a big step forward by creating a "closed loop" of production planning and control. The closed loop means that the results of what has been planned on the shop floor — the actual production completions — are fed back into the planning system and replanned. This practice enables production planning and control to be very reactive when things change on the shop floor. Instead of producing idealized plans, the planning system can take into account real problems that occur within a production plant.

Most MRP II systems can also integrate production planning and control systems with the customer order-entry, material-procurement, and financial-accounting systems. In this way the entire management of the organization can be controlled through one set of integrated programs. The MRP II system becomes the principal method of managing the production organization.

The better MRP II systems available as "off-the-shelf" packages can also integrate an organization across more than one physical location. This systemization provides the opportunity to control an entire enterprise through one integrated computer system. In addition, these systems allow MRP and CRP concepts to be applied to distribution logistics within a multiplant/multisite environment. Such an approach is known as distribution requirements planning (DRP).

Prerequisites of Success with MRP II

Manufacturing resource planning has been implemented in thousands of companies throughout the United States, Europe, and the Far East. There has been wide divergence in the success these companies have achieved — because MRP II will be successful only if the organization using it is committed to an MRP II approach and is willing to put a great amount of effort into the system's implementation and maintenance.

For MRP II to be successful, company managers must abandon former methods of controlling the operation, particularly the manual expediting tasks, and rely entirely on the MRP II system to plan and control production. A high level of commitment from senior managers is essential.

MRP II also requires a high degree of discipline and accuracy. Inventory figures must be accurate, bills of material must be accurate, production routings must be accurate, and the master production schedule must be well constructed. In addition, data entry must be timely, accurate, and consistent, including such tasks as purchase receipts, production completions, stock adjustments, and customer-order shipments. Of course, it can be argued that to be successful, any system — manual or computerized — requires such levels of accuracy and consistency. This is true. However, the visibility and sensitivity of an MRP II system quickly results in wrong calculations and misleading recommendations. As a result, the system itself is frequently blamed for the problems, and people return to the old manual methods of control.

User education is another prerequisite of a successful MRP II implementation. MRP II systems tend to be complex, and people using the system should be thoroughly trained in its use. This training is time-consuming and expensive; many companies attempt to introduce MRP II without spending enough time, money, or effort in education.

However, a more subtle prerequisite is that the system must not be considered the property of the MIS (or management information system) department. To make it work, operations managers, supervisors, and operators must take ownership of the system. Too many companies perceive of MRP II as a new computer system. In fact, it is an entirely new way to manage and control the business.[4] The tool is a computer system — the changes are management issues.

Why MRP II Has Not Created Agile Manufacturing

There is no denying that manufacturing resource planning has provided enormous benefits to thousands of Western companies. These benefits have manifested themselves in improved customer service, reduced inventories, reduced production costs, and greater flexibility. However, the introduction of MRP II has not been the resounding success expected by its proponents. A recent letter in *Information Week* magazine stated that "the claims of MRP II consultants seem to approach fraudulence at times."[5] Be that as it may, MRP II has not resulted in the flowering of a new age of productivity in Western manufacturing.

Several reasons can be cited for its lack of success. One is that expectations have been set too high. One survey of MRP II implementations showed that only 58 percent met expectations.[6] Another reason is that mistakes have been made in how the systems have been implemented. Principally, however, MRP II implementations fail to match expectations because company managers lack the vision and perseverance to make the changes required to obtain the benefits. They view its implementation as a new computer system rather as part of the introduction of radically better business methods.

Pundits and consultants were enthusiastic about MRP II in the late 1970s because they saw the potential for fundamental change in the management of the manufacturing industry. With these new systems, Western manufacturers have a powerful new productivity tool. They can now do the following:

- obtain up-to-the-minute information
- match production to customer orders
- schedule delivery of materials precisely when needed
- measure shop-floor success in meeting production schedules
- analyze the production process to initiate improvement
- balance material flow through the plant
- eliminate shop-floor queues
- significantly reduce inventories of both finished products and materials
- significantly reduce production cycle times
- significantly improve customer satisfaction

When a company using MRP II is able to tap the full potential of improvement, the results are dramatic — but full potential is difficult to achieve. MRP II requires a level of commitment to long-term change that is often lacking.

Let's not paint a gloomy picture. The fact that very few companies achieved the full potential of an MRP II approach does not mean that the use of its systems is unsuccessful. The vast majority of companies implementing MRP II did achieve improvements in all the important areas of their business. The United States, where the majority of MRP II systems have been implemented, has the highest level of productivity in the world. Productivity in countries like France and Britain, where MRP II systems gained momentum during the 1980s, has increased dramatically.[7] But MRP II did not create a revolution in productivity and customer service — that revolution took place in Japan and was led by companies with very different ideas.

MRP II and Agile Manufacturing

Companies successfully applying agile manufacturing techniques are those prepared to make fundamental changes in the way they do business. Agile concepts were introduced gradually over a 40-year period by Japanese companies like Toyota Motor Company. Continuous improvement and total quality are the cornerstones upon which radical changes are applied to all aspects of the company's business. The successful companies in Japan and other Pacific Rim countries recognized the need for a fundamentally different approach as they sought to rebuild their industries after World War II. Western companies that have most successfully introduced agile-manufacturing techniques are those that have seen their markets severely eroded by foreign competition. In many cases, radical changes have been made in response to a threat to the company's very existence.

Many Japanese pioneers of agile methods simplified production processes so that computerized planning and control systems were unnecessary. These innovative companies then introduced manual control methods like the *kanban* system. As time passes, these companies have begun to use computerized systems without resorting to Western-style MRP II.

Other successful Japanese companies have taken the opposite approach. They have recognized the value of MRP II, providing it is combined with the radical changes in management philosophy required to bring about total quality and continuous improvement. A comparison of the Nissan Motor Company and Toyota is interesting in this regard.[8] Nissan has fully embraced MRP II techniques, whereas Toyota has deliberately shunned them and has developed manual methods.

Results from the two companies are very similar. Both have excellent records for quality, innovation, and customer satisfaction; low levels of inventory; high degrees of production flexibility; and reduced time-to-market. Nissan is proud ot its record: 100 percent on-time deliveries, 99 percent achievement of its production schedule, lead time reduced from sixteen days to six days, and component stock turns better than 300 times per year.

The key issue here is that agile manufacturing's success has very little to do with the computer systems the company chooses to adopt. Success has much more to do with the leaders and innovators within the company having the vision and determination to become a world-class company. This concept is equally true within the United States where some successful agile manufacturers have approached the quest for excellence in different ways. Xerox Corporation took the approach of competitive benchmarking as the tool through which innovative and radical change was made. Motorola stressed quality through its 6-Sigma program, which has been highly successful in motivating this technology-led company to become a world-beater in semiconductor development and production. Harley-Davidson took an approach similar to that of Toyota and introduced quality programs and just-in-time (JIT) manufacturing using a kanban-style production flow. Northern Telecom centered its efforts around the ideas of time — short cycle times, short time-to-market, timeliness of deliveries, and the elimination of wasted time.

These companies have all achieved their spectacular improvements differently. The one thing they have in common is that they each have been through the painful process of reconstructing from the inside out. Each recognized that radical change was needed in order to have a chance of beating the competition on the global market. Gone were the days of 10 percent yearly improvement rates. One hundred percent improvement was needed — and needed fast. Companies like these have risen to the challenge and succeeded.

Operations Research and Production Management

MRP II and other modern production and distribution techniques were developed in a very different environment from those of world-class manufacturing. Although the objectives of a Western approach to production and inventory control are consistent with agile manufacturing methods, the theoretical framework within which each was developed differs. Consequently, some essential differences require discussion.

Most of this book is taken up with examining the software needs of a company making the journey away from traditional manufacturing. Implementing an agile approach requires radically changing many different areas of the company's business, and these improvements require that significant changes be made to the computer systems controlling production. In a Western context, these changes can be built most effectively upon the solid foundation of MRP II, which is fully compatible with the ideas of agile manufacturing.

The basics of modern Western production and inventory management were developed in the United States at a time when computer systems were beginning to become widely available to industrial companies. Simultaneously, colleges and universities were engaged in important research into the application of mathematical techniques to the solution of business problems. This theoretical work became widely accepted as ideas of operations research were adopted within U.S. companies.

The premise of operations research in solving business problems is that if the problem can be modeled mathematically, then the issues relating to that problem can be analyzed and an optimum solution can be calculated. This approach can be widely used within manufacturing and distribution industries. Such issues as demand forecasting, calculation of safety stocks in relation to planned customer-service levels, transportation optimization, economic lot sizes, and shop-floor scheduling are areas where ideas of operations research can be readily applied.

During the 1970s and early 1980s thousands of companies employed these techniques in an attempt to develop a "rational and scientific" view of business problems. Many projects were successful and achieved the desired results. However, some inherent consequences of this approach run counter to agile ideas. These consequences include the following:

- complexity
- modeling of "reality" that inhibits change
- an assumption that less-than-perfect trade-offs are required
- inflexibility

Mathematical models of business problems tend to be complex. Issues dealt with are complex and the mathematical techniques are sophisticated. The result is that the people using the analysis frequently do not understand where the numbers have come from and therefore are unable to validate information or apply their own judgment to the task. In turn, users either unthinkingly follow the computer's lead or manually override it — thereby making many of the analysis figures useless. People working this way often feel alienated because the only person who can understand the system is some mathematical boffin working in the MIS department. This situation invariably leads to conflict.

Agile manufacturers stress simplicity of production methods and the empowerment of shop-floor operators and supervisors. Much of the authority and responsibility for running the plant or warehouse lies with the people on the shop floor. There are no layers of middle managers to make decisions and provide liaison between departments; the people doing the job are responsible for their own planning, scheduling, and quality control. For this to work, the systems must be simple and transparent to those using them, and these people must be trained to use them effectively. In general, the operations-research concepts are not compatible with this new approach to the management of operations and clerical personnel.

A second aspect of the operations-research approach is the idea of modeling reality. A mathematical model of a business situation (like a production plant) is created so that the problems and issues can be analyzed and resolved. If this is the case, then the model has to consider countless variables and conditions. Indeed, modeling this one aspect alone can be complex. Operations-research engineers have been amazingly clever and creative in the development of these models; they have been able to include every aspect of the problem on hand. The unfortunate consequence is that the model, when completed, includes aspects of the production plant that require change or elimination. Things that should be eliminated can easily be considered as accepted and inevitable parts of the process.

For agile companies striving to improve processes and procedures, continuous improvement is essential. Changing something once it is "baked" into the system is difficult, particularly when people using the system do not fully understand the consequences of each aspect of the model. For continuous improvement to work and for employee suggestions and problem-resolution processes to be effective, there must be both a clear understanding of the processes and an assumption that anything can be changed and improved.

An example of this within MRP II and traditional cost accounting is the idea of "standard cost." In most companies, the standard cost of a product or subassembly is calculated taking account of scrap rates for components. This calculation builds into costs an allowance for quality problems and component defects. If supervisors receive material variance reports showing that their work centers are producing at standard cost, then they think they are doing well. However, by agile standards they are doing poorly because they are scrapping material. While the model shows an acceptable performance, a detailed analysis is required if the supervisor is to realize that the department is performing poorly.

The next issue, the assumption of trade-offs, betrays a basic philosophy that is contrary to world-class thinking. Agile manufacturers always strive for perfection. They are practical enough to know that they have not yet achieved it and that each step along the road is likely to be more difficult than the previous step. Still they pursue their endeavor. Operations research is always looking to balance contradictory variables. When there are many variables to contend with, complex regression analysis is used to gradually home in on the most acceptable compromise. This practice is good mathematics — but poor business management.

When agile techniques are first introduced into a plant, there is always an argument about whether the outlandish expectations can be achieved. Is zero inventory practical? Is 100 percent quality achievable? Can suppliers really provide JIT deliveries? One key task of leaders is to show that these things are achievable. However, this can only occur by throwing away old ideas and re-inventing the company. These changes take time. Continuous improvement means gradual and consistent change, with each step getting closer to the goal. While the goals are high indeed, they almost certainly do not conform to the model.

An arch example of this kind of thinking is the general acceptance of economic order quantities (EOQ). EOQ is a trade-off between the cost of carrying inventory and the cost of setting up to make one batch of the product. EOQ takes into account spurious factors such as storage costs, order-placement costs, and setup costs. An agile manufacturer would say that the only thing to know about batch sizes is that they should be halved in 90 days and then halved again. The goal is a batch size of one; genuine just-in-time manufacturing. It is rarely necessary to drop to batch sizes of one, but the process of continuous improvement will head that way.

Finally, flexibility is another key aspect — including flexibility of product mix, product volume, production methods, and the people manufacturing the products. Agile companies also look for flexibility throughout the design process and in the introduction of new products and enhancements. An operations-research approach invariably leads to inflexibility because it conceives of the model with a specific approach in mind and then drives the production plant in accordance with that model. This inflexibility hampers companies dealing with some significant areas of agile manufacturing.

Traditional Production Control and Operations Research

Developed from an operations-research viewpoint, the majority of the production planning and control techniques used by Western manufacturers suffer from some of the shortcomings discussed in the previous paragraphs. These techniques become stumbling blocks within companies endeavoring to move into agile manufacturing because they are complex, inhibit change, make sub-optimum trade-offs, and do not provide the flexibility required in an agile work environment.

MRP II systems are complex. As previously stated, they require considerable training and education of the people who use them, and they very often become the domain of the MIS department instead of being used as the primary tool for controlling the business. MRP II systems inhibit change because they have built into them fundamental concepts relating to production planning and control within a production plant. For example, most systems require the use of some kind of work order with detailed reporting of job-step completions on the shop floor. At some stage in the implementation of agile methods, the work-order approach (or at least detailed reporting) will need to be eliminated, making the traditional MRP II system a stumbling block. Many companies have found ways to work around this kind of problem and have achieved great success despite the shortcomings of the systems used to support their operations. However, it would be better if the systems themselves could be vehicles of improvement instead of problems to be resolved.

The move toward a more flexible production approach makes traditional systems less helpful. The flexibility of having production operators cross-trained so they can move from one work cell to another invalidates the capacity requirements planning. Short cycle times and customer lead times require the shop-floor people to plan in a much more dynamic way than is envisioned by traditional MRP II systems. Changes in product mix and production volumes require world-class companies to take a different approach to production scheduling.

From a more positive standpoint, the simplification of the production process brought about by introducing agile manufacturing methods results in less complex systems being required. Instead of designing a system to model the complexities of a traditional production environment, the agile company reduces the complexity so that simpler, more easily understood systems can be used. Simplifying the processes within a production plant is extremely difficult. A key task of company members implementing the new methods is to study their production methods and to simplify the process.

Many agile companies have been highly successful using MRP II systems because their basic objectives are consistent with each other. However, there are some inherent shortcomings in the MRP II philosophy and other production and distribution systems that make them a hindrance to an agile approach.

Characteristics of Agile Software

Software needs of agile manufacturers vary greatly because these companies tailor their operations to customer needs, their own industry, and the global marketplace. In addition, while agile concepts are common to all of them, the emphasis placed upon each aspect differs from one company to another.

As noted earlier, some very successful U.S. companies have implemented agile manufacturing in different ways. For example, Motorola emphasizes quality, Xerox emphasizes competitive benchmarking, and Northern Telecom concentrates on time. These various approaches require different kinds of software. Unlike the MRP II path, it is wrong to consider that there is a single correct approach.

Still, some characteristics of software systems are essential to their success within an agile environment. While system details may vary substantially, the principles are the same. These common characteristics include integration, simplicity, flexibility, openness, and accessibility.

Total Integration

A key emphasis of agile manufacturing is the elimination of waste. Waste is defined as any activity that increases costs but does not add value to the product or service the customer receives. Closely tied to this idea is the elimination of unnecessary administrative work, including the accessing and processing of information. An integrated system has all the information needed to run the business on a single data base with a single point of entry for that information.

Many traditional companies have computer systems that have evolved over time. Some systems may be "home grown"; others may be standard packages. Some systems may run on one kind of computer and others on another. This scenario leads to disintegration. For example, it is common for order entry and customer service to be a separate system run on a separate machine from the manufacturing systems. Similarly, the financial and accounting systems may be handled on yet another machine.

A number of problems are created by disintegrated systems. First, there is the complexity of having different systems, different technical environments, and programs that work differently. Second, important information needed by one system is not available because it is held on another system (for example, when the order-entry system does not have any manufacturing schedule information available). Order-entry people are unable to make firm delivery promises because the availability of product must be verified and authorized by production-planning personnel. This procedure introduces delay, confusion, and considerably more effort and clerical work.

Some companies attempt to overcome this problem by interfacing the two systems. The data needed by one system from another system is fed across each day (or each week) as a batch update. This method is useful but results in the information being out-of-date and held in two places at the same time. A basic rule of software development is that information should be held in only one place. Holding it in two places takes up unnecessary space and makes it easy for the two pieces of information to get out of sync.

Other companies with disintegrated systems do not attempt to bring them together. The various systems are run independently and any common information is entered twice. This method gives the worst of both worlds; not only is the information stored twice and out-of-sync, but there is also double work performing data entry twice.

A company with integrated systems has the advantage that all information is available in one place. This minimizes the effort of entering data, keeping information up-to-date and accurate, and (if the systems are well designed) providing a similar "look and feel" for all aspects of the system. Someone using the order-entry system can quickly become familiar with accounts receivable, for example, because the methods of using the systems are consistent.

The most powerful aspect of an integrated system is the availability of information. All information about production, inventory, customer orders, quality, and so forth is readily available to anyone within the company (providing they have appropriate authorization). There is no delay, no extensive paperwork, no wasted time and effort; accurate, up-to-date information can be accessed easily. The availability of accurate and timely information can be regarded as a quality issue. If the information is accurate, the company is better able to provide quality service to customers.

Another aspect of an integrated system is the integration of the system's various features. The features must be designed to work together (for example, integrating work-order and nonwork-order schedules with production planning). As a company progresses with agile manufacturing, production-planning people will want to eliminate work orders from the planning and execution process. This change

usually cannot be done all at once. Changes are best introduced gradually over time, starting perhaps with one seg-
ment of the production process. Consequently, production
planning, master scheduling, and MRP systems must be able
to fully integrate work-order and nonwork-order schedules.
It must be possible within the system to manufacture the
same product at the same time within the same factory using
a work order or a rate-based schedule. This level of integration allows for a smooth and low-risk change from traditional
work orders to rate-based scheduling.

Even after rate-based scheduling has been fully implemented and production planning simplified, many companies still find a need for work orders. Mixed-mode
manufacturing, where the plant has some production controlled by work orders and others planned using rate-based
schedules (or no schedules at all), is common because there is
often a need for special projects, new product introductions,
or other production requiring detailed planning and tracking.
These production methods can continue to be controlled
using work orders even when the majority of the plant is controlled in simpler ways. The systems must be able to handle
these situations through an integration of both the information and the functionality.

Simplicity

An important aspect of a JIT or agile approach to
manufacturing is the application of simplicity. Many Japanese
books and training classes on manufacturing methods stress
the idea because simplicity brings with it clarity and understanding. If systems and processes are simple, then people
can understand them. If people understand processes and
objectives, they are better able to contribute to company
goals. If processes and systems are complex, people will have
difficulty understanding what they have to do and why they

do it. Consequently, they are unable to recognize problems as they arise and solve the problems.

The application of simplicity in a production or distribution environment has provided enormous benefit to thousands of companies. The detailed study and analysis required to fully understand the processes and then devise methods of simplification create an innovative atmosphere within which the organization can make radical improvements. The continuous-improvement process must also center around the idea of simplification. As operators within the plant bring forward ideas and suggestions, often meeting in a quality-circle format, the attention of the entire work force is on the simplification and improvement of the process.

Note that there is a difference between a system being "simple" and a system being "simplistic." A simplistic system is inadequate for controlling the business because it does not handle nuances of operations. On the other hand, a simple system considers all issues involved in a modern and sophisticated production plant, but does so in a way that is transparent to people using the system. People are not intimidated by a simple system; they regard it as a useful tool, even a friend.

A fine example of simplification can be seen in the rise of Federal Express Corporation. The idea was simple: to provide reliable overnight delivery of packages. The mechanism was simple: to bring all packages into a central hub (Memphis, Tennessee) at the end of the day and air-freight them to their destination. Traditional transport economic theorists considered this approach too simplistic and unworkable. In fact, when Frederick W. Smith, the founder of Federal Express, developed the concept and wrote it up as part of a college graduation project, he received a low grade. While the idea was simple, working it out required considerable management and leadership skills.

One key issue is simplification of the production process. Computer systems cannot be simple and effective if production processes are complex and convoluted. Agile manufacturers place great emphasis on simplifying the production process. Professor Richard J. Schonberger, a pioneer in this area, subtitled his world-class manufacturing book "The Lessons of Simplicity Applied."[9] Making things complicated is easy, but making a process simple takes effort, perseverance, and intelligence. The application of agile techniques has the effect of simplifying production planning and control processes because many elements of complexity and confusion are removed. For example, manufacturing products through production cells instead of work centers significantly reduces the number of production stations, thus simplifying the process and procedures.

Reducing inventory simplifies stockroom control because it is easier to handle less material. Without inventory, stockroom control would not be an issue. Reducing cycle times simplifies the process because the product is made more quickly and because operators can understand and participate in the entire process of manufacture. A traditional production process in which components are made into subassemblies that are inventoried, then turned into finished products, and finally packed ready for dispatch is complex and time-consuming, creates high inventories, and is difficult for operators to fully understand.

Long cycle times provide a double set of problems because they inherently result in high work-in-process (WIP) inventory. If WIP inventories are high, then it becomes necessary to keep track of and account for that inventory. This situation forces companies to use cost-accounting and inventory-control systems that require procedures, transactions, and effort. If cycle times are reduced to the point where WIP inventories are very low, complex accounting and control systems become unnecessary. Materials move quickly through the production plant and inventory value is low. Therefore, there is no need to pedantically track receipts, issues, and adjustments of these materials. Agile companies try hard to reduce cycle times, and this is good in itself. However, the "knock-on" effect of short cycle times significantly reduces system complexity because the material value can be deemed immaterial by the accountants.

Designing Simple Systems

Another aspect of simplicity is the software itself. Many things can be done to make software simple and easy to use. Most of these issues will be addressed as each aspect of software for the agile manufacturer is examined. However, some general aspects of software design lend themselves to simplicity from the user's point of view.

The first aspect is consistency. Most manufacturing and distribution software systems have grown over time and many people have contributed to their design and programming. Each contributor's style and approach results in the system operating differently as the user moves from one module to another. Consistency in look, feel, and features of software contribute greatly to its simplicity of use.

This problem of consistency has been addressed by the development of industry standards for "user interfaces." These include SAA from IBM, the Macintosh and Windows graphic approach, and Motif from ANSI. While these standards are relatively new and not yet in wide usage, they will become the norm as time goes by.

Another problem with conventional software is redundant data on screens and reports. A standard package aiming to meet the needs of a wide range of manufacturing and distribution companies with different processes and procedures, of necessity, must include features and functions not used by every company. Invariably, the screen and reports show unneeded or irrelevant data elements. Cluttered screens are confusing and detract from the system's simplicity and ease-of-use. There is a balance to be struck here because a well-designed system will contain carefully thought-out features and functions which, while not currently needed by the company, will be valuable in the future. Removing features by taking extraneous data from the screens and reports can lead to a lack of flexibility when business conditions change. Also, if standard packaged software is being used, the removal of unnecessary data can be a tedious, time-consuming programming task that has to be repeated each time a new software version is released.

There is room for some compromise in this aspect of software design, but busy screens lead to confusion. Simplification of screens and reports adds to the simplicity and usability of the system. If people find a system easy to use and fully understand its features and functions, they will be willing and able to use the system effectively.

A third aspect of software design that leads to simplicity is a logical flow. People will forgive shortcomings of functionality if the system has a fundamentally logical and understandable flow. The important element here is that the system must be designed so that people can understand where each feature fits into the overall business scheme. Software systems can be frightening. A system with a flow of features that is well thought through greatly reduces that fear and distrust.

When MRP II and other systems taking the operations-research approach were being widely introduced into Western manufacturing industries, it was considered that adding features and functions improved a system. The best systems were the ones with the most "bells and whistles." In recent years this view has changed. Simpler systems, which do not burden the user with all kinds of subtle new features, are easier to use and thus more effective. Well-designed simple systems are more useful than systems with all manner of detailed complexities. The detailed complexities contained within some systems have been added due to a need for the features by one user. These features have been added to ensure that all aspects of production or distribution are fully modeled within the system. As agile manufacturing techniques are introduced and production processes simplified, there is no longer a need for additional confusing features. A clean, sharp system that addresses the real issues of production planning and control is preferable to a system with multitudes of additional features and functions.

Flexibility

One of the few things we know about the future is that things will change. This axiom is particularly true for an agile manufacturer. In addition to being subject to changes taking place within the industrial world and within its own industry, a manufacturer also instigates internal change through the use of a continuous-improvement approach. Software designed to be used by a company undergoing transition must be flexible. If the system forces the company to operate in one particular way, then it will hinder the continuous-improvement process. The system must be adaptable to changing needs of the company over time.

The idea of continuous improvement is that there are no quick fixes. Change and improvement must come by vigilantly and consistently discovering ways to improve design, manufacturing, distribution, and marketing operations. Everyone in the company is charged with the task of improving processes and operations. Techniques like quality circles focus attention onto an aspect that can be improved, and people in that area are given the right and responsibility to make improvements. Radical change comes about by applying hundreds of small improvements.

Usually a company cannot implement the new techniques in one spasm of activity. The changes, too radical to occur all at once, must be introduced gradually and implemented differently in various parts of the company. One plant approaches changes differently from another plant. The needs of one division may be quite different from that of another division. Introducing agile manufacturing in each area will determine how supporting software will be used.

For example, many traditional companies plan master schedules on a monthly basis. As agile techniques are introduced, the master-scheduling period may be reduced from monthly to weekly, then to biweekly, and finally to daily. Many production-planning systems are "bucketed." In a bucketed system, planning periods (months, weeks, etc.) are established within the system and all requirements and replenishments used for production planning are placed in appropriate time buckets. A company using a system that has monthly buckets will find it very difficult to improve the master-scheduling process because the master-scheduling system is forcing a monthly approach. A more flexible system will have "bucket-less" master scheduling in which requirements are calculated daily or by shift and results are aggregated into time buckets at the last minute when the reports and inquiries are run. This kind of system allows for the use of variable-length time buckets and provides flexibility as improvements are introduced. In addition, because time buckets used in different areas of the company can vary, changes can be introduced differently in different parts of the company. A system that is useful to an agile company must have this kind of flexibility.

A designer of manufacturing and distribution software for world-class organizations faces a dichotomy between the need for simplicity and the need for flexibility. Flexibility requires a selection of available features and functions so that appropriate avenues can be chosen according to situation needs. However, the need for simplicity would suggest restricting the functionality of the software by removing features not currently in use. This area of software design requires a great deal of creativity and intelligence. Simplicity and flexible functionality can be combined successfully. Doing so is the challenge for agile software designers.

Agile companies face such dichotomies in many areas. Traditional wisdom says the following:

- Good customer service requires high inventory.
- Large batch sizes are needed for economical manufacture.
- High quality requires systematic inspection.

Each of these fundamental principles has been proven wrong by agile manufacturers who have done the detailed work, provided the vision, and overcome the problems in order to bring themselves to world-class status. The same is true with software; flexibility and simplicity can be combined with intelligent design, an understanding of the needs of the users, and a vision for world-class levels of manufacturing and distribution.

We have discussed these issues in terms of the design for manufacturing and distribution software. However, most companies do not design their own software — they purchase standard packaged software. The various packages must offer flexibility. As the company journeys toward agility, manufacturing and distribution methods will change significantly.

The software must be able to reflect those changes. Better yet, the software must incorporate tools to aid the transition and facilitate the improvement process.

Openness

Software used to support a company in transition must lend itself to interfacing with other systems and to networking between multiple computers. It is increasingly important that computer systems be able to communicate with each other. Some agile aspects are time-dependent activities requiring real-time interaction between the various software and systems controlling operations.

The increasing use of personal computers (PCs), work stations, and local processing necessitates the free access of information throughout the company. This level can be achieved if hardware and software can be effectively interconnected. Until quite recently, every computer system had a large central machine where the computer processing was done. The results of this processing was passed down to the people through reports and inquiries. One trend of the 1990s is the movement toward distributed processing whereby the software operates locally throughout the company, providing computer power to people on the shop floor, in the warehouse, and in the offices. This style of computing is very much in line with a world-class approach because it allows the responsibility and authority for production planning, inventory control, production scheduling, and quality to be placed onto the people where they work.

An example of this approach is the scheduling and control of production cells. Instead of having all production cells controlled from a central planning system, each cell can have its own small system allowing people in the cell to perform their own planning and control locally. These cells are then linked together and into a larger machine that handles the central operations.

Agile manufacturing software must be capable of supporting a multisite, multiplant configuration. At first glance, two agile aspects appear contradictory. One is the movement toward local control and autonomy. The best people to plan and schedule production are those responsible for achieving production. The best people to work together for continuous improvement are those who do the jobs every day. The best people to be responsible for quality are those making the products because they control the quality of the process. Therefore, the people who do the work are given the authority and responsibility (as well as the relevant education and training) to control their own areas.

The second supposedly contradictory aspect is the increasing integration of operations. The design department becomes directly linked to the production floor; suppliers are electronically linked with people placing orders, often by electronic data interchange (EDI) and by electronic funds transfers (EFT). The JIT approach to inventory control requires close liaison between various parties that contribute to the production process. In a multisite environment this structure can include a complex network of information flow from customer-service and order-entry people, various production plants, plants that feed fabricated parts and subassemblies, and vendors supplying raw materials.

These two aspects appear to be contradictory because one requires local control and the other requires central control. Yet, these two approaches can be combined successfully through computer systems that are linked together effectively.

Until recently, most hardware vendors have attempted to use proprietary protocols for linking together systems so that customers would continue to require hardware from that particular vendor. Only in the last few years have hardware manufacturers seen the need to get together and establish standards that allow a greater degree of interconnectivity for their customers across multiple hardware platforms. This significant change in the industry has wide repercussions for both users and suppliers. Customers now can more easily establish a total system using various kinds of equipment that all easily link together. This results in greater commonality between computer hardware, and the machines themselves will become more similar. In turn, this commonality will reduce the price of computer hardware still further because the machines themselves will become less a specialized piece of technology and more a standard product.

Manufacturers of computers and other systems hardware have recognized the need to have various kinds of equipment to communicate effectively. In recent years, standards bodies have been established, attempting to create standard protocols for the interchange of information. Some of these standards have become widely used. MAP standards, for example, are used to communicate between shop-floor devices like robots and numerically controlled machines. ANSI standards in the United States and EDIFACT standards in Europe are now used extensively for the EDI transfer of information between trading partners.

These changes in the way hardware systems are configured make it even more important that the software be able to operate in distributed, heterogeneous environments. The skill of the system designer is to make full use of the sophisticated system's configuration while at the same time making the system simple and relevant for users.

Accessibility

Information contained within the computer systems must be easily accessible to the system users. All computer systems have reports and inquiries that provide information in standard ways and formats. These reports are the basis for information retrieval. However, most companies find that because they require specific information that is not readily available on standard reports and inquiries, new systems must be written to provide data that addresses those needs. The dynamics of agile manufacturers enhance this need. Information on the data base must be readily accessible to everyone within the company.

World-class performance measures often differ greatly from those of traditional manufacturers (see Chapter Eight). In addition, those performance measures change over time and from one plant to another. In particular, traditional management-accounting techniques with variance reports and overhead absorption calculations are replaced by new nonfinancial methods of performance measurement that require reports usually not available within a standard manufacturing and distribution software system.

Plant personnel also need to extract information on an ad-hoc basis for analysis purposes. Typical analysis includes setup times, distance moved through the plant, scrap analysis, and number of warehouse picks. Precise reporting needs will vary considerably between plants because their needs and problems differ. If the only reports available are the standard reports within the system, then the information required is not accessible to users.

A traditional approach to this problem is to have the MIS staff write new programs to extract the information required and present it according to user specification. Two problems arise with this. First, the MIS departments in most companies are frequently overloaded with projects, and writing a new management report is usually low on their list of priorities. Second, once the new report is programmed, it can only be changed by going back to MIS and by repeating the process of requesting time and resources. Many of these reports are required for a single analysis and are unneeded on an ongoing basis.

Data-base information can be made accessible to people using the system through an easy-to-use report writer. A report writer is a software tool designed to allow managers, supervisors, and specialists within various departments to write their own reports and inquiries and to access information from the computer systems without the need for specialist programming. These tools are called fourth-generation languages (4GL) and they can be readily understood and used by people with a minimum of training.

The report writer gives access to information held within the system to the people who need that information to run the business. They can produce their own reports, both standard and ad hoc. They can do their own analysis and are not restricted by the system or the schedules of the MIS department. Other useful features of the report writer include the ability to upload and download data between machines. For example, financial information can be downloaded from the central data base onto a PC for the production of graphics or presentation materials. Similarly, information held on a local processor, such as quality statistics or word-processing files, can be easily uploaded to a central machine. Many 4GLs provide additional features that enable users to update files, create screens, and develop simple subsystems. This kind of advanced tool puts additional flexibility into the hands of the user without the user's having to learn programming techniques.

For a 4GL report writer to be easy to use, information held within the data base must be understandable to the user. Data-base definitions and dictionaries in conventional production-control systems often are cryptic and difficult to understand. When the information required is held on more than one file in the data base, it is necessary to know how the files are related before the information can be accessed. These problems can be overcome by the 4GL having its own dictionary designed to clarify these kinds of problems. The 4GL dictionary, which will have been populated automatically from the system's standard data base, contains descriptive and cross-reference data, making information in the data base readily understandable. The dictionary will also have standard headings so that users do not have to enter report headings every time, standard validation criteria so the report writer can be more vigilant in its error checking, and standard field positions and subtotal criteria so that the user does not need to enter complex formulae.

The problem of how to link files together can be solved by creating "views." A view is a piece of information within the 4GL data dictionary that shows how a series of files are linked together. Views created by the MIS department or the software manufacturer are given easily understood titles like "sales analysis data" and contain all the files required to provide information relating to the title. Users do not need to know or understand how the views are put together; they only need to know that the view contains all the information they require for particular tasks. In reality, the views can never cover 100 percent of user requirements, and some users do need to learn more of the detailed structure of the system. However, the majority of day-to-day requirements can be met without the users having any knowledge of the data base.

Another useful feature of a 4GL data dictionary is the addition of "virtual fields." A virtual field contains a calculation from information held on the data base and is shown as a separate file. An example would be average sales for a product. The data base contains actual sales for the last twelve months, and a virtual field is defined as the total sales divided by 12. While the virtual field does not exist on the data base and does not take up space on the files, its value is calculated everytime the user accesses the virtual field. Virtual fields reduce the amount of work the user has to do when creating a report or inquiry because many repetitive calculations already have been done within the data dictionary.

The use of report writers has some drawbacks. Report writers are sophisticated pieces of software, and their use tends to be a heavy load on the computer. When there are many departments in a company and all of them are using the report writer to provide their information, the computer can become overloaded and processing time can be excessively long. Often the response time for on-line entry (for example, of sales orders) noticeably deteriorates when 4GL reports are being used. Another drawback is that many departments within the company will have similar needs, and it does not make sense to have a person in each department writing reports that are similar to those of another department. When the 4GL also contains file updating and subsystem development tools, great care is required to prevent damage to the data base. This aspect must be controlled by the MIS department. There needs to be a level of central coordination of the use of report writers and 4GLs to avoid duplication, to optimize machine usage and response time, and to ensure the integrity of information on the data base.

Summary

Providing software support for a mature agile manufacturer is relatively straightforward. This is because much attention already will have been given to simplifying production and distribution processes that, in turn, simplify software requirements. However, providing software for manufacturing companies in transition is a much more subtle and complex task.

The traditional MRP II approach to manufacturing systems often hinders the introduction of world-class methodologies. While MRP II and agile objectives are entirely congruent, MRP II contains a rigid operations-research approach to solving production and inventory problems that needs changing if the systems are to be useful in an agile-manufacturing environment.

Although software-support needs vary considerably from one company to another, some underlying principles of software design essential for agile support include the following elements:

- *Integration*
  Systems must be fully integrated so that information is entered only once, is up-to-date and accurate, and is accessible to users.
- *Simplicity*
  Programs, screens, and reports must be designed in a way that makes them simple and easy to use. Simplification is a key agile aspect and the software systems should reflect this.
- *Flexibility*
  Throughout implemention of agile manufacturing techniques, a company will make significant changes to business operations. Software must have the flexibility to allow users to introduce new techniques in one area while retaining the old approach in other areas. This flexibility lends itself to a smooth low-risk approach to agile implementation.
- *Openness*
  Supporting software must lend itself to easily interfacing with other systems and networking between multiple computers. The interfaces typically will

of real-time quality control, computer-aided design/computer-aided manufacturing (CAD/CAM, shop-floor data collection, and automated ware-housing systems.

- *Accessibility*
  Data-base information must be readily accessible to users of the system. The information is required for creating performance measures and ad-hoc analysis reports, and for interfacing to other systems. A well-designed 4GL report writer provides this access.

CHAPTER 2

Agile Manufacturing

While lacking clear or formal definitions, the term "agile manufacturing" is one of those irritating pieces of jargon that has grown up around manufacturing industry of late. However, the term is useful and does capture the importance of changes currently taking place in the manufacturing industry. To be able to compete in world markets requires far more than just implementing a few new techniques. Companies must transform operations so that their products, quality, and service are second to none. Agile manufacturing is not a set of techniques — it represents a total change in the way manufacturers approach their business.

Changes in Manufacturing Industry

The concept of a manufacturing industry is relatively new. Prior to the industrial revolution, goods were manufactured by craftsmen in small private workshops employing either an individual or a small group of skilled people making products one at a time. Things were made that way for five thousand years until the industrial revolution brought large numbers of workers off the farms and into the burgeoning cities to labor in factories. This revolution occurred in Europe during the eighteenth century and a little later in the United States. These changes were brought about by the brilliant thinking of a few men who developed the concepts of making products from interchangeable parts and who invented the machines and processes of mass production.

All of these changes revolutionized Western society by changing the nature of work, changing the demographic landscape of countries, altering family life by creating jobs external to the family home and farm, and by creating unheard-of prosperity. There were also significant negative repercussions. The abject poverty and degradation of people in the industrial cities was vividly portrayed by Charles Dickens in novels like *Oliver Twist*. Frederick Engel's descriptions of living conditions in working-class Liverpool, England, in the early nineteenth century shocked the nation and spurred on his friend Karl Marx, who was at that time developing entirely new economic theories. Their *Communist Manifesto* called for the workers of the world to unite and "throw off your chains" and inspired the likes of Lenin, Trotsky, and Mao.

New Management Methods

Throughout the nineteenth century, as industrial development became more and more sophisticated, new organizational methods were required to manage the emerging companies. This change fostered the idea of professional management of industrial organizations. Railroads, the telegraph (and later the telephone), and improved roads opened the possibility of diverse organizations controlled from a central location. Brilliant new techniques were developed by companies like Sears Roebuck, DuPont, the steel industry, and the railroads to maintain control and productivity within these more complex corporations.

The "Scientific Management" movement, developed in the late nineteenth century, emphasized the time study of industrial processes on the theory that the "right" way to do any task will result in maximum productivity. The introduction of mass production lines by Henry Ford revolutionized production technology, and the conglomeration of smaller companies into a single central giant corporation (like General Motors in the United States or Imperial Chemical Industries in Britain) revolutionized industrial management. These concepts of professional management were legitimized in the late 1920s and 1930s when the great universities and learning centers such as Harvard developed business schools to teach these dynamic new techniques. The introduction of professional associations followed to codify and regulate these new approaches.

The Technology Revolution

Scientific and technology revolutions of recent years have again changed forever the landscape of modern industrial life. Production methods have changed beyond recognition, entirely new materials and technologies are being introduced every day, and the labor content of products has fallen dramatically. We are far more dependent upon machines and plants than ever before. The rate of change of technology, with its resultant "future shock," has transformed our approach to working life.[1] A recent survey concluded that the average worker in the United States needs to be retrained every five to seven years, rendering traditional skills a disadvantage because they must be unlearned. The rate of change is rapidly increasing, and the major challenge we face is our ability to harness that change and adapt to it.

Throughout the 1950s and 1960s the United States dominated manufacturing industry. The economies and manufacturing industries of Japan and Western Europe had been destroyed in World War II. The aftermath also initiated social changes in European countries as ideas of democratic socialism were put into practice. Government regulation or ownership of industrial enterprises, nationwide welfare and health-care systems, government sponsorship of industrial research and education, and the rise of labor movements introduced cultural changes that significantly impacted industrial life. The United States found itself on top of the industrial heap. Its rich resources, educated work force, superior technology, and vibrant manufacturing industry far outstripped the rest of the world. From the early 1950s through the late 1970s America was successful in every area — from new products to productivity, from research and development to exports. No one could touch U.S. industry.

Japanese Competitors

Unfortunately, an element of complacency and arrogance crept into many American companies, and the strides being made by European and — more particularly — Japanese companies went unrecognized. During the 1970s and 1980s foreign competition became a major factor in the United States. Entire industries (consumer electronics, motorcycles, photographic equipment, for example) were wrested out of the hands of major U.S. corporations by vital and innovative foreign companies. The U.S. automotive industry was impacted significantly by the influx of Japanese products that were better made, better designed, more in touch with the market, and less expensive. The automobile giants that had dominated not only the U.S. car market but also much of the international market were left reeling as their market share and profitability fell dramatically.

Many theories have been proposed to account for the success of Japanese companies over the last 25 years, and several factors have contributed to their success. These factors include a government-planned economic development program with low-rate long-term loans, a low-wage work force that was prepared to sacrifice in order for companies to prosper, and an awareness of designing to meet customer needs. One crucial aspect of Japanese success in the Western marketplace has been innovative new production techniques. These techniques include an emphasis on quality, just-in-time (JIT) production control, employee involvement, and production flexibility. These techniques have been given a number of different names and even more acronyms, but agile manufacturing best describes the series of techniques and approaches used by good Japanese companies to provide customers with high-quality, well-designed, reliable products that have swept the world markets in diverse industries from ship-building to consumer electronics.

A sad aspect of this scene is the fact that the majority of the ideas and techniques which made these Japanese corporations successful had ironically been developed in the United States. The crucial use of innovative production and quality-control techniques was taught to Japanese companies by people like Dr. W. Edwards Deming and Joseph M. Juran in the 1950s. U.S. corporations at that time did not acknowledge or foresee the importance of a "Total Quality Control" (TQC) approach to their manufacturing business. The concepts of JIT manufacturing, to a large extent pioneered by Henry Ford in the 1920s, had not been carried through into postwar production technology.

The challenge for all Western manufacturing companies at present is to implement aspects of agile manufacturing that will enable them to compete on the world market. It has become eminently clear that no industry is exempt from the onslaught of competitive pressure from companies (usually foreign companies) that are able to present better service, quality, design, and flexibility.

There is room for some optimism here. Many Western companies have recognized the competitive challenge and have been able make the radical changes required to meet and beat that challenge. Some of these companies are well-known, large, multinational corporations and others are small, privately held businesses. Size and complexity is of little relevance to a company's ability to innovate and succeed in the turbulent tides of 1990s manufacturing. In some ways larger corporations have a more difficult task than smaller organizations because they have to cut through years of entrenched bureaucracy in order to make the changes needed. Smaller companies are often quicker and better able to respond to competitive challenge.

Agile Manufacturing

Agile manufacturing is a useful heading under which to place a series of techniques that have been employed by good companies to bring about unprecedented improvements in quality, productivity, and customer service. These techniques are not new, many have been available for several decades, and others have been developed gradually over the last 30 years by innovative companies like Toyota. What is new about these techniques is that, when they are implemented in a concerted way, they bring about enormous improvements in products, responsiveness to customers, innovation, and flexibility. These improvements give companies such an edge in the marketplace that they are able to take a commanding competitive lead.

It is not the purpose of this book to discuss in detail the various elements of agile manufacturing.[2] However, the following introduction to each important aspect will put into context the discussion of the software needs of companies implementing these new techniques. We will view it from four angles to show its wide-reaching changes. Agile manufacturing does not so much represent a series of techniques as a fundamental change in management philosophy.

The sequence chosen for our discussion is significant:

1. quality
2. just-in-time manufacturing
3. people
4. flexibility

This list represents that amount of interest and involvement observed among Western manufacturers in recent years. Most companies have got the message on quality. Everyone knows that to compete in world markets requires levels of quality that far exceed previous expectations, and most companies have some kind of active quality-improvement program in place. Many companies have embraced at least some aspects of JIT manufacturing, and some have a fully developed JIT implementation program addressing the issues of cycle times, batch sizes, vendor relations, and synchronized manufacturing. Just a few companies are making any significant changes in the way people are managed within the organization. Such changes are difficult because the issues deal with deeply held cultural understandings, and several years of consistent effort is required to make changes of this sort. Very few Western companies are seriously addressing flexibility as a weapon in the battle for world markets. There are some notable exceptions, but the average Western manufacturer seems unaware of the issue.

Quality

In agile manufacturing, quality is a way of life. The approach taken to quality is quite different from that of traditional manufacturers because agile manufacturing is concerned not only with detecting quality problems but also resolving the source of those problems. A traditional manufacturer, recognizing that there will always be rejects with any manufactured item, will often build acceptable reject levels into production plans and customer orders. An agile manufacturer recognizes no such "reality" and sets the quality goal for zero defects or 100 percent quality. This level of quality is achieved by a systematic, long-term program of identifying quality problems and harnessing the entire work force to resolving those problems.

A traditional manufacturer employs a large staff of inspectors whose job is to check the quality from all materials purchased from vendors and everything manufactured within the plant. Subassemblies are inspected at each stage in the production process to ensure that they "pass." The thinking behind this approach has always been to ensure quality by having the product inspected carefully at each stage in the production process by independent, trained inspectors rather than by the operators making the goods.

This approach is not only expensive but also ineffective. Many companies have found that improved quality does not come from having more inspectors. The production staff does not feel that product quality is its responsibility if its work is inspected by others throughout the process. The inspection department is considered to be responsible for product quality. This situation is untenable for all concerned. Inspectors become thoroughly frustrated because they are held responsible for something over which they have no control. An inspector can do nothing to improve product quality; all he or she can do is find faults after the event. Production operators are equally frustrated because they have people looking over their shoulders all the time checking up on them and assigning blame when things go wrong.

An agile manufacturer places the responsibility for quality onto the people who do the job. A pride of ownership is fostered on the shop floor because people have both the authority and responsibility for quality. Craftsmen have always taken pride in items they make because they feel the products in some way represent themselves. Much of this outlook has been lost as modern industry developed from craft-based to factory-based production. Agile manufacturers can engender this kind of pride and personal responsibility within the production work force and beyond. For this approach to be successful, people responsible for quality must have the tools, training, and authority to do what it takes to ensure zero rejects. A number of techniques that lend themselves to this approach are widely used. These techniques include statistical process control (SPC), the posting of quality results on the shop floor as they occur, quality circles where quality problems can be analyzed and resolved, and appropriate education and quality standards that emphasize the importance of quality to the company.

The quest for quality does not stop at the shop floor. Every department of the company is affected by the need for zero defects. Because the design of the product often has more bearing on quality than the production process, many agile manufacturers change the way products are designed to incorporate concepts of quality. The change requires a much closer link between design engineers and the people who make the products. Some companies, including some of the very best Japanese companies, have developed a radically different design approach. People from all aspects of the business are brought together into a team to work with design engineers in the development of new products. The purpose of these teams is to design products the customers want, to produce products at a price they are willing to pay, to achieve 100 percent quality, and to bring innovative ideas to market quickly. This ambitious objective has been highly successful. Sales people, marketing people, design engineers, production engineers, cost accountants, field-service people, shop-floor people, and administrative staff are brought together to work as a dynamic "tiger team" for developing a new product.

The results have been phenomenal. New products have been introduced in a fraction of the time, product quality becomes radically better, the design is "right the first time" because the range of skills contributing to the design ensures that all aspects are fully considered, and the company introduces products customers want to buy.

JIT Manufacturing

Just-in-time — a cornerstone of agile manufacturing — is another term that has become common parlance in manufacturing circles over the last few years. JIT is concerned with the elimination of waste, where waste is defined as any activity that increases costs without adding value to the product being manufactured or the service being provided. Examples of waste are inspection, unnecessary movement of materials, shop-floor queues, rework or repair, storage of inventory, and overhead personnel.

The objective of JIT manufacturing is to change the production process to eliminate waste. A lot of emphasis is placed on eliminating inventory, not only because inventory is an expensive waste in itself, but also because high levels of inventory have traditionally been used to hide other problems. Vendor delivery problems is an example of this syndrome. It has been accepted practice for vendors to occasionally deliver materials later than scheduled. A traditional manufacturer will hold additional safety stock of those materials so that a late delivery will not disrupt production. This inventory method promotes waste and is a prime target for a JIT approach.

Although an ill-defined term, JIT usually incorporates at least the following approaches:

- Change the shop-floor layout to reduce movement of materials.

- Reduce production setup times so that products can be made in very small batches (ideally a lot size of one).
- Synchronize the manufacturing process so that sub-assemblies and components are available just when they are needed and not before (or after).
- Create mutually beneficial relationships with suppliers.

Shop-Floor Layout

A traditional manufacturing plant is usually not laid out for JIT manufacturing; rather it divides the factory into machine centers that group together machines of similar type. Jobs are processed by moving the semicompleted material from one machine center to another according to the production routing of the product. In a typical traditional manufacturing plant, actual run time is frequently less than 10 percent of total production time. The rest of the time is taken up by queuing and material handling.

This approach requires that the product be moved physically a considerable distance during production as it is transferred from one work center to another. Movement is not only wasteful in itself, but also leads to material damage, high inventory investment, and long production cycle times. A third problem is that this layout leads to complex control procedures. When a production route is complex and the job takes several days to complete, complex planning and control systems are necessary to enable managers and supervisors to keep track of production schedules.

Just-in-time layout minimizes the movement of materials, people, and tooling. This system usually involves laying out the shop floor as a series of manufacturing cells. In a cell system, dissimilar machines are grouped together based upon the process to be performed within the cell. Instead of a batch of materials being moved between several work centers, a number of different machines are grouped into a cell so that all production tasks can be completed with virtually no movement of materials.

Such a grouping of activities assumes a degree of repetition. Techniques of group technology are employed, where products are categorized according to production processes. Cells are dedicated to the production of a group of products and contain all machines, assembly areas, and test equipment necessary to make that specific group of products.

In addition to eliminating unnecessary material movement, this layout style lends itself to flexibility. The rate of flow through a cell can be changed easily as customer orders change. People within the cell are trained to perform multiple production tasks and can be assigned to a wide range of activities based upon current needs. A cell can also become a responsibility center where production personnel control their own quality, schedules, material planning, and job assignments.

Setup-Time Reduction

Setup-time reduction — a major goal in its own right — provides benefits in lot-size reduction, shorter lead times, and improved flexibility. Reduction of batch sizes usually requires a reduction of setup times. A just-in-time manufacturer spends a great deal of time and effort studying setup times and devising methods of reducing them.

Traditionally, large production batches cause high levels of inventory — both in work-in-process (WIP) and finished goods. Batch sizes (or lot sizes) frequently are determined by setup times. If eight hours are needed to set up a machine, then making a large quantity after doing all that setup work is generally desirable.

In contrast, a prime objective of JIT manufacturing is to reduce inventory. In fact, inventory reduction is a passion in many JIT plants not only because inventory is costly and wasteful in itself, but also because high inventory levels traditionally have been used to cover a multitude of other shortcomings. High levels of buffer stocks are used to alleviate problems resulting from such situations as bad scheduling, poor vendor delivery, poor quality, and inaccurate record keeping. An important tool for the reduction of inventory is reduction of production batch sizes, and an important tool for reduction of batch sizes is reduction of setup time.

Setup time reduction cannot be achieved by one or two changes. Reductions are achieved by a concerted and systematic effort throughout the entire production process. The chief offenders are identified first — those processes with the longest setup times and largest batch sizes. Experience proves that setup times can be reduced by approximately 50 percent by merely studying the process and by establishing better procedures. Setup processes can be studied by examining current written procedures, by discussing setup with the people who do the tasks, and by videotaping the process in action and analyzing the activities step-by-step. Once initial studies have been completed and changes implemented, setup time can be further reduced by modifying dies, tools, and jigs. Techniques and innovative ideas associated with these programs have names like single-minute exchange of dies (SMED) and one-touch changeovers (OTC).

In some cases, achieving the level of setup time reduction required for just-in-time manufacturing while using existing equipment is impossible. New production equipment must be purchased specifically for the purpose of reducing setup times throughout the entire production process. Typical of obsolete or nonadaptive equipment are heat-treatment ovens, wave soldering baths, extrusion or paint processing requiring lengthy wash-downs, and some specialized test equipment.

Setup time reduction can have a significant effect upon work-in-process inventories, cycle times, and production flexibility. Benefits are achieved through many small improvements over an extended period, and meaningful analysis and monitoring methods are needed.

Synchronized Manufacturing

The ideal production plant will set production capacity for each work cell so that each operation is synchronized with operations that come before and after. Unfortunately, there probably are no ideal production plants, but JIT manufacturers strive toward the goal of totally synchronized production in which each work cell receives exactly what it needs when it needs it.

Several problems result from operations lacking in synchronization. Large WIP inventories are possible even when batch sizes are small and setup times short. Shop-floor queues are usually the result of bottleneck operations, poor production planning, and part shortages.

Bottleneck (or limiting) operations are those with low capacity and large loads. These operations may be considered the weak link in the chain of production processes that go into making the finished product. For example, a bicycle manufacturer might be able to make 5,000 wheels per day, but can only make 2,000 frames; if the manufacturer continues to make 5,000 wheels each day, WIP inventories will build up because another stage in the process can only cope with 4,000 wheels (assuming there is no requirement for spare wheels).

Better synchronization can be achieved by approaching the problem from two different perspectives: production planning and shop-floor execution. Production can be planned so there is near-perfect synchronization between loads on each work cell. Such perfection can be achieved by careful master scheduling and detailed capacity planning through limiting work centers. Some available software attempts to optimize production schedules by automatic optimization of bottleneck work centers. This software is limited in its applicability and has yet to find general acceptance.

Most JIT manufacturers divorce production planning from shop-floor execution and use some kind of "inventory-pull" method on the shop floor. A shop-floor pull system merely means that each supplying work center does not make anything until the next work center requests it to do so. The method is based on the Japanese-developed *kanban* system. Subassemblies and components are supplied to the work center only when that work center issues kanban cards to the supplying work center. This approach can yield excellent synchronization, flexibility, and minimum inventories, providing each work center has a short setup time, small lot sizes, and short cycle times.

The closer a production plant comes to synchronized production control, the nearer it will be to genuine just-in-time manufacturing. Such an achievement requires excellence in every area of the production process.

Vendor Relationships

A traditional manufacturer tends to have an adversarial relationship with vendors. Procurement personnel are judged primarily on purchase price variance; in other words, they are rewarded for getting the lowest price. The company will ensure that there are several suppliers for each part so that these vendors can be pitted against each other when contracts are negotiated and to ensure continuity of supply.

In contrast, agile manufacturers employ mutually beneficial vendor relationships. Vendors are selected primarily on the basis of quality, delivery reliability, and flexibility. The theory is that the benefits of reliability, consistent quality, and just-in-time deliveries are far more valuable than a few pennies off the price.

Techniques required for a good vendor relationship include single sourcing, certification, and openness. Single sourcing uses just one supplier for any component or raw material that is purchased and thereby reduces the total number of vendors. The idea is simple — a supplier doing a lot of business with your company will treat you with more care. If you have large orders going to fewer suppliers, then you are able to create closer, more cooperative relationships with each supplier. This approach has been used very successfully by companies such as Xerox, which reduced the vendor base from 5,000 to below 200 as a part of its successful JIT/Quality program.

Plunging into a single-sourcing arrangement will be irresponsible if vendor performance records are not known. Performance is certified by a process that assesses every supplier's record of quality, delivery reliability, and flexibility. Deliveries from a certified vendor will be received without the need for incoming inspection and will often be delivered directly to the shop floor.

Vendors are certified to supply particular components when performance in each aspect is up to standard. If a vendor does not match company standards, then the buyer will seek to resolve the problems with the supplier. As the certification process develops, many vendors will be dropped and the total number will be reduced. Eventually the company will purchase components from only a small number of certified suppliers.

Unfortunately, vendor certification recently has gained a negative connotation because some larger companies have intimidated smaller suppliers into providing concessions on service and price. Many large companies have used their clout to force smaller vendors to hold large inventories of finished components in order to support their customers' just-in-time requirements. Such practices only push wasteful safety stocks back onto the vendor rather than allowing the vendor to move into a world-class arena. This approach runs contrary to agile vendor relationships, where mutually beneficial long-term relationships are the primary aim.

A good starting point for improved relationships is openness. Agile manufacturers have an open-door policy with vendors. There is a free exchange of technical information, clear communications of needs, expectations, and performance, and the sharing of financial information about products being supplied. This open communication provides a basis upon which trust can be built. As these relationships progress, further cooperation can develop. Many companies find that suppliers can assist with the design of components and subassemblies and that their contribution results in better designs, higher quality, and lower costs.

Agile manufacturers need unprecedented levels of quality and reliability, just-in-time deliveries, and short lead times and often require additional services such as deliveries in sequence or special point-of-use packaging. In return, the vendor receives large orders, longer-term contracts, and fair prices. A vendor will often require just-in-time payment for just-in-time deliveries. Thus, a happy circle frequently develops as suppliers themselves move into agile manufacturing. The techniques required to provide the superior levels of quality and service have the effect of making the supplier much more productive, and product price can be reduced. In fact, many companies are paying less for components and raw materials while receiving JIT deliveries and zero defects.

People Management

Frequently, the most difficult problems to solve when introducing agile techniques relate to the management of people. A company committed to agile manufacturing must be prepared for fundamental change in management style and philosophy. After years of mutual distrust and poor relationships, both managers and production personnel often find it difficult to make the required changes.

Many plants have discovered that shop-floor operators, appropriately managed, can make the largest contribution to the attainment of agile goals. Technological changes and automation have taken much of the physical work out of the production process; agile manufacturers attempt to make full use of their work force's hands-on knowledge of how to make products better, faster, and safer.

New approaches to the management of people can take several months or even years to introduce. What is required is trust, honesty, and success. Developing trusting relationships between people and managers takes time and experience.

Transfer of Responsibility

A traditional manufacturer tends to treat production personnel as suppliers of labor — not suppliers of thought. Middle managers and supervisors are employed to make day-to-day decisions, resolve problems, and take responsibility for production. An agile manufacturer will take great steps to give operators greater control of their daily work. They will have prime responsibility for product quality, for scheduling preventive maintenance, and for attaining production targets.

This change does not mean that middle managers, engineers, and specialists are no longer required; however, their roles change. Instead of being primary decision makers, they become expert advisors, coaches, and mentors to shop-floor operators and are the people who become the champions of change to senior managers.

If operators are to have responsibility for production quality on the shop floor, then they must have the authority to stop the production line when quality problems arise. They must have expert assistance readily available. They must have the equipment and training to enable them to perform these new tasks. The end result is better-managed and better-controlled plants. Instead of a few inspectors and planners trying to look after the entire operation, now hundreds of people see that their responsibilities include quality, delivery to schedule, and customer service.

Education and Cross-Training

Agile manufacturers dedicate a great deal of time and money to educating employees. Some of this effort is devoted to ongoing training in JIT manufacturing, quality control, and customer service. Another important area is the training of operators to do a wide range of tasks within the production plant.

This cross-training of people allows manufacturers to have greater flexibility because operators can be moved between tasks. If customer demand is high within one product group and lower in another, then people can be moved to an area where they are most needed. If product mix is such that the production rate on different work centers has to change so that synchronization is maintained, then the number of people in each work cell can be changed.

Problem Solving and Quality Circles

The introduction of quality circles has been received with mixed opinions in the West. In the early days of JIT, many experiments with quality circles did not work well, often because they were not accompanied by other important aspects of agile manufacturing.

The purpose of a quality-circle approach is to have every employee involved in solving production (and other) problems. No matter how this participation is organized, whether through quality circles as such or other methods, the goal is to have the active participation of all employees, each of whom can contribute his or her own individual skills and experiences.

In the past, problems were solved by middle managers, engineers, and specialists. In an agile plant, the entire work force is involved in one or more projects aimed at continually improving products, processes, and services. These programs have found spectacular success in many companies because an atmosphere of teamwork and common cause has enabled people who previously had very little opportunity to contribute to become innovative and resourceful problem solvers. In addition, the people involved in these efforts enjoy their work more because they have a wider variety of tasks and because they and their ideas are treated with respect.

Flexibility

While Western manufacturers are working hard to attain world-class status in terms of quality, delivery performance, and customer service, many top Japanese companies are building upon existing excellence. A major challenge facing Western manufacturers in the near future will be one of flexibility.

At one time the belief was that techniques of just-in-time manufacturing were suited only to repetitive production processes with highly predictable schedules. While it is true that JIT is easier in such an environment, many companies are now extending just-in-time concepts into more complicated production situations where product quantities and production mix change frequently.

In their 1988 report "Flexibility: The Next Competitive Battle," Professors Jeffrey G. Miller, Jinchiro Nakane, and Arnoud De Meyer show that many leading Japanese manufacturers are beginning to emphasize production flexibility and price as their future competitive edge.[3] They are striving to make products at a low cost and with a high degree of flexibility to meet customer needs.

Two aspects to flexibility are important: production flexibility and design flexibility. Production flexibility is achieved when the company can offer short lead times, when product mix can be changed significantly from day to day, and when people are cross-trained to manufacture a wider range of products. A company that can offer this level of flexibility (without a price penalty) has a significant competitive advantage.

Design flexibility is related to the company's ability to introduce new products and modifications to existing products. Market needs change quickly. A company must understand current and future customer needs, must develop innovative products, and must get those products to the marketplace quickly. Once, seven to ten years were needed to bring an automobile from conception to production; today, Nissan can introduce an entirely new car in less than three years. Recently, Chrysler has been able to achieve similar levels of new product introduction. This level of flexibility to meet customer needs in both the short term and the longer term is the mark of a company truly committed to agile manufacturing.

Some Success Stories

A thousand success stories surround the implementation of agile manufacturing techniques in Western companies. Those presented here exemplify corporations that are stable, well documented, and where all aspects of agile manufacturing have been implemented.

Xerox Corporation

Xerox Corporation was an enormously successful organization throughout the 1960s and 1970s. It achieved substantial growth each year by exploiting the innovative techniques of plain-paper copying so successfully that the word "xerox" entered the English language as a verb. ("I will xerox that for you.") In the late 1970s, however, Xerox began to see the emergence of Japanese competition as several small companies began making inroads into the copier market with small, high-quality, inexpensive copying machines. This competition was not recognized as a major threat by Xerox because the competition products were at the low end of the market, competing in the low-margin sector that Xerox felt it could afford to loose.

Unfortunately, in two or three years these Japanese companies became a recognized threat to Xerox. They began to produce mid-range copiers, Xerox's bread-and-butter business — the kind found in every office. While the highest profits were made on high-volume copiers used by professional printing organizations, its biggest market was for the mid-range office copiers. Xerox had previously owned greater than 80 percent of this market. The wolf was at the door, and Xerox had to act. The initial analysis was very discouraging to Xerox managers. The copiers being made by the Japanese companies were not only less expensive than Xerox machines, but they were also of a much higher quality. The competition delivered higher-quality design, features, reliability, aesthetics, and customer satisfaction. Competitive benchmarking showed that one Japanese competitor could *sell* a product for less than Xerox could *make* an equivalent machine.

Xerox rose to the challenge. In the troubled days of the early 1980s, under the leadership of David Kearns, Xerox radically changed its approach to business. Management implemented just-in-time manufacturing techniques and introduced a companywide quality program that impacted the entire staff from design to sales to support and administration. Xerox pioneered the use of competitive benchmarking as a tool for focusing where change is required and for monitoring the success of that change.

Xerox had notable success with vendor relationships. The vendor base was substantially reduced and many cooperative partnership arrangements have been made between Xerox and its vendors. Another area of particular success was the education program designed to bring the entire corporation into one mind on the issues of quality. From 1982 to 1987 all 100,000 Xerox employees received 48 hours of education entitled "Leadership Through Quality," which emphasized customer expectations and zero defects.

The result of these changes — very painful for those involved and especially for people whose jobs and lives were affected — is that Xerox has now gradually regained command of the market and is once again the premier plain-paper copier company in the world. Xerox has also recognized that the market is changing and has embarked on a number of new ventures within its perception of document processing.

Harley-Davidson

Harley-Davidson presents another success story of an American manufacturing company able to transform itself in the nick of time. On several occasions over the last ten years Harley-Davidson teetered on the brink of collapse. The problem lay in the fact that Harley had such a commanding lead in the U.S. heavy-motorcycle market that the company became complacent. Products were selling so well that they saw no need for innovative new designs. There was a waiting list of customers to buy Harley-Davidson motorcycles, and the company priced the machines according to its own needs rather than improving productivity to give customers better value. Harley had such brand loyalty that some customers even had their bodies tattooed with the Harley logo.

Competition in the marketplace almost annihilated the entire Western motorcycle industry during the 1970s. There are now no motorcycle manufacturers in Britain, none in France, only one in Germany, and only Harley-Davidson in the United States. Companies like Honda and Kawasaki entered the market with small, inexpensive machines, established a foothold, and then began to produce large machines that directly competed with Harley-Davidson. Harley struggled for many years to stay competitive. Compared to the new Japanese machines, the Harley product was old-fashioned. Yet no investment money was allotted to redesign and retool the product.

The engineering of Harley machines had not changed in years. In contrast, the new machines from Honda used advanced design techniques, particularly in the power train, that Harley could not match. Harley quality was renowned for being patchy, so much so that a typical Harley owner needed a considerable amount of knowledge to keep the machine on the road. The ground beneath a Harley was also patchy because the company had been unable to prevent the bikes from leaking oil. In fact, oil leaks were a Harley trademark, a basic quality problem that the company had never appropriately addressed.

Harley-Davidson even ignored customer tastes and requirements. For several years Harley riders had been stripping and embellishing machines with fashionable features making them "easy riders," touring bikes, and hot rods. Harley had never catered to this more specialized aspect of the market despite the fact that these Harley owners were their most loyal and influential customers. Not until Honda began to produce machines to attract this focused market did Harley recognize a need to provide specific products for this important sector of their customer base.

The story of Harley-Davidson during the 1980s is one of high risk and hard work. The company was purchased by a consortium of employees in 1981. These committed people steered the company through several years of near disaster, several times pulling the company from the brink of collapse. In 1982 Harley-Davidson petitioned the U.S. government for temporary protection against foreign competition under section 201 of the 1974 Trade Act. This protection only lasted five years, but it was enough of a breathing space to allow Harley to introduce agile manufacturing techniques and customer-driven design and marketing. Those painful days resulted in Harley-Davidson becoming one of the most productive companies in America, bringing innovative new products to the marketplace, and improving quality beyond anyone's expectations. In other words, Harley employees transformed their organization into a truly world-class company through the implementation of TQC, JIT manufacturing techniques, and a customer-oriented approach. Now one of America's most successful companies, Harley has branched into several different areas of manufacturing in order to buffer the company from the peaks and troughs of the motorcycle industry. It has developed innovative management-accounting techniques and received several prestigious awards including one from former President Ronald Reagan.

Stadt Industries

At the other end of the scale, when it comes to successful implementation of just-in-time and agile manufacturing,

I think of Stadt Industries Pty. Ltd. Stadt is a small Australian manufacturer of heating and climate-control equipment for domestic and industrial use.⁴ As they began losing ground to more customer-oriented competitors, company managers began to worry. Their immediate solution to the problem was to move to a larger facility, increase capacity, and purchase expensive numerically controlled machines. This approach had some benefit; quality improved and Stadt Industries could meet increased customer demand. However, these moves did not entirely solve the problem.

Customer feedback showed that while product price, quality, and service were acceptable, the lack of immediate availability of products often lost the customer to other, sometimes more expensive, suppliers. A study showed that Stadt had two primary problems: shortage of parts and long lead times. Parts shortage was caused by poor planning and control; long lead times were caused by large batch sizes and long production runs. If a customer ordered a product, several weeks might lapse before that kind of product was scheduled for manufacture. When it finally was manufactured, the finished product would be delayed due to shortage of components. This situation is not atypical for small manufacturing companies making the transition from low-volume to high-volume production.

Aided by some professional consultants, the company took the classic JIT approach. Cause-and-effect analysis was done using fishbone charts and Pareto charts to highlight problem areas, and extensive training was put into place for all employees. Additional one-on-one education was also required for production operators who were unfamiliar with and suspicious of JIT manufacturing ideas. The next step was to implement a more formal inventory-control system. Previously, the company had used an informal approach to the storage of parts. Consequently, finding the right parts was often difficult, stock levels were high, and shortages were common. The new system was straightforward — single fixed locations were assigned to each item and strict controls instituted to ensure accuracy of transactions. Where possible, stocking locations were situated near assembly areas that used the part. This change resulted in better stock accuracy, less damage, and reduced material movement.

The introduction of mixed-model manufacturing, the first major challenge, required reducing setup times so that the product could be manufactured more frequently in smaller batches. This alteration reduced wait time for a product when its production run was not scheduled for some weeks and also resulted in a significant increase in volume output owing to better scheduling and better real utilization of production facilities. It also resulted in increased parts shortages because the company did not have inventory control procedures in place to allow for the flexibility of demand resulting from mixed-model scheduling. The solution to this problem was the use of kanbans.

Kanbans were introduced gradually, starting with a few critical components and then spreading out to all manufacturing components and subassemblies. Kanban techniques were then extended to include purchased items and, with the cooperation of vendors, delivery lead times were dramatically reduced.

With these techniques in place, Stadt embarked on a formal continuous-improvement (or *kaizen*) approach. Employees were trained in the ideas of continuous improvement and, once the program gained momentum, an enormous number of improvements were made in quality, flexibility, cycle times, vendor relationships, and product design. The benefits have been substantial. Inventory has been reduced by 60 percent, work-in-process cut to a maximum of 40 hours, shortages virtually eliminated, and production flexibility increased. However, the most important result has been the increase in customer satisfaction. Today Stadt outstrips many competitors on flexibility and responsiveness.

Aston Martin

Aston Martin is a much-loved name among British sports-car enthusiasts.[5] Their luxury sports cars have been classics for over 70 years; they are expensive, powerful, and classy. What is not generally known is that the first completely new Aston Martin in 20 years, the Virage, was introduced just two years ago. The Virage was an instant success, achieved critical acclaim, and created long order books for the company. This success was matched with the introduction of the Volante, a convertible, in 1991.

Despite the success of its vehicles, Aston Martin was hit badly by the recession. It cut the work force by 16 percent in March 1991 and reduced production from six to 4.5 cars per week. With a history of financial and production problems, the company has rarely been profitable, and ownership has changed hands several times. Currently, the company is 75 percent-owned by Ford of Europe with the balance of stock being held by the company's chairman and another investor.

The company was in trouble. Despite a fine reputation and successful new products, managers were very worried about long-term survival. Cutting costs and improving productivity was critical. A review of Aston Martin's production processes revealed the typical symptoms of a traditional Western manufacturer: high inventories, shortages, delayed schedules, high work-in-process, large batch sizes, confused and cumbersome systems. Senior management quickly committed itself to making the changes required to implement agile manufacturing methods. The managing director stated that radical change was needed to ensure survival and began looking for a breakthrough into greatly reduced costs, improved quality, faster response to marketplace needs, and then into profitability. Setting a time line of two years to get the job done, company managers established the "Breakthrough Recovery Plan" presented in Table 2-1.

This approach to the implementation of agile techniques was started at Aston Martin in 1991. While it is still too early to realistically assess its success, the programs are working well and improvements are beginning to manifest. With any implementation of JIT or agile manufacturing approaches, the collective improvements that take place over time are those that provide dramatic results; no one feature can be pinpointed as the one that provides the benefit. For the program to be successful, the whole approach must be orchestrated in a systematic manner as Aston Martin has done. In addition, the change of approach radically affects the way people work and the way they are managed. Managers must prove that they are serious about improvement and that they really will start treating their employees as assets. In industries that have suffered from years of industrial strife, to create trusting relationships between the company and the working people takes a long time.

Applied Digital Data Systems

Applied Digital Data Systems, Incorporated, (ADDS) a division of NCR Corporation (now owned by AT&T), manufactures computer terminals in Hauppauge, Long Island, New York. In the early 1980s ADDS moved all manufacturing operations off-shore to Taiwan and Korea in order to reduce costs and improve profits. In 1990, ADDS brought its manufacturing back home and built a total-quality JIT production

Aston Martin's Breakthrough Recovery Plan

1. **Vision and Leadership**
   Recognize that the impetus for radical change must come from the company's top management.

2. **People**
   Break down the traditional management-worker barriers that plague much of British industry, institute education, and training programs. Develop a companywide atmosphere of employee involvement.

3. **Design Engineering**
   Open design engineering to a wider team approach. Introduce concepts of concurrent engineering and quality through design.

4. **Production Engineering**
   Acknowledge that people are the backbone of the changes required in shop-floor procedures. Make changes in floor layout, setup-time reduction, batch-size reduction, cellular manufacturing, and materials control.

5. **Purchasing**
   Introduce world-class purchasing including supplier reduction, supplier certification, quality at source, and JIT deliveries.

6. **Inventory**
   Reduce inventory through kanban control and JIT purchasing and manufacture.

7. **Productive Capability**
   Use employee involvement and problem-solving teams, improve material flow, improve working practices to increase productivity significantly, increase production capacity, and reduce overtime requirements.

8. **Quality**
   Improve "first time" quality and significantly reduce the cost of quality through a "Total Quality Control" program.

9. **Financial and Performance Measures**
   Redesign methods of management accounting and performance measurement to bring them into line with a world-class manufacturing approach.

Table 2-1. Aston Martin's "Breakthrough Recovery Plan"

operation in Long Island. The new operation is ten times more productive than it was in the early 1980s, quality is second to none, and lead times to customers are better than ever.

The point of decision for ADDS came in 1987 when the company suffered massive losses and NCR considered selling it. A core group of managers persuaded NCR to back them on a dramatic company overhaul. The new strategy was the quick delivery of quality, customized products to customers. The approach required abandoning past manufacturing processes and creating a new company culture of quality and service.

ADDS employed a three-pronged strategy: quality, just-in-time, and people involvement. Quality issues were addressed through quality education to all employees, implementing quality approaches to design, production, and vendors, and giving authority and responsibility to the production people. JIT techniques implemented included fast setups, small batches, flexible manufacturing that makes to customer order, zero inventories, and the now classic U-shaped cellular shop-floor layout. The plant produces 700 to 800 terminals each day and, according to George Munson, the director of manufacturing, "That can be 800 different products or the same product. We are flexible enough to handle it either way."

The software supporting the facility has been tailored to meet flexibility and quality needs of the operation. All production is triggered from a customer order, and the shop-floor control "Flex" system tracks the manufacture and shipping of the order through the entire process. The Flex system makes extensive use of bar-coding so that operators can quickly access product information. Because ADDS makes over 6,000 different products, operators need to know the configuration, customization, and component information. Owing to product variety, the Flex system keeps track of which components are used on each product and checks back to the engineering data to prevent errors. Flex also provides information that is used for continuous improvement, real-time status reporting, quality tracking, and failure analysis.

An interesting caveat to this case study is that the company's MIS department hindered implementation of these radical new ideas. Staff members did not participate in the implementation, saying they were too busy running the old MRP and inventory systems. As a result, the Flex system was designed by the operations managers and programmed by a member of the quality department.

To date, the company has achieved the following:

- Manufacturing overheads at ADDS have dropped from 18 percent to 8 percent of throughput. Inventory has been reduced by 70 percent despite an enormous increase in the number of finished products.
- Nonproduction overheads have been cut by two-thirds.
- They now ship custom-made products in 24 hours.

And ADDS workers proudly affix an American flag label to every unit they ship.

These and countless other success stories demonstrate that the concepts of agile manufacturing are not only applicable to Western manufacturing, but can turn companies from the point of disaster to being market leaders on a world-class scale. Changes are not easy to make. They require a thorough restructuring of the company, complete rethinking of design, manufacturing, and marketing approaches, and often a total change of management philosophy. These changes are difficult, painful, and require significant management skill to be successful. But the reality is that for a company to survive and prosper in the 1990s, such changes must be made. A company not competitive on the world market cannot survive as we move into an era of increased globalization.

Summary

How the manufacturing industry has developed from the days of small craft workshops through the industrial revolution and into the technological age of the twentieth century is a fascinating story of innovation, leadership, brilliant thinking, and sheer hard work. For the three decades after World War II the United States was head and shoulders above all other countries in manufacturing technology, productivity, and new-product development. During the late 1970s and early 1980s companies from Japan and other Pacific Rim countries began to make significant inroads into the Western manufacturing base, and these companies are now the dominant force in several industries.

Countless Western companies, recognizing that international competition could jeopardize their business and even their survival, have adopted new methods of manufacturing and distribution that are known collectively as "agile manufacturing." Its purpose is to bring about the radical changes required for a company to be competitive on the world market and usually encompasses the following aspects:

- A new approach to quality aimed at meeting customer requirements and providing zero defects.
- Just-in-time (JIT) production techniques including new methods of shop-floor layout to minimize material movement, setup-time reduction to reduce lot sizes,

synchronized manufacturing to provide an even flow of production through the plant, and a new approach to vendor relationships resulting in single sourcing and certification of suppliers.

- Significant changes in the management of people aimed at focusing all employees on quality improvement, waste elimination, continuous improvement, and production flexibility. These changes include the transfer of responsibility for quality and production to the shop-floor operators, substantial education and cross-training of personnel, and development of employee-involvement programs for problem resolution and suggestions.
- The development of more flexible production methods and rapid introduction of new products to meet the needs of the marketplace.

Thousands of Western companies are successfully applying agile ideas to their operations. These companies have seen radical improvement in their competitiveness on the global market. In some cases the introduction of a world-class approach to manufacturing has enabled a company, on the brink of disaster, to survive the onslaught of international competition. The techniques, largely developed in Japan, have been adapted successfully to a Western environment, and many innovative ideas have been introduced to give Western companies a competitive edge.

CHAPTER THREE

———

Cellular Manufacturing, Rate-based Scheduling, and Capacity Planning

*T*he introduction of agile manufacturing techniques radically changes shop-floor control, production planning and scheduling, and inventory control. Primary aims in this area are the elimination of waste, reduction of production cycle times, and zero inventories. Many techniques are used to achieve these goals, and these techniques must be implemented together if the goals are to be achieved. Furthermore, do not expect to enact a series of sweeping changes and achieve world-class status overnight. These changes take time, commitment, and perseverance.

Agile manufacturing requires a total change in the way a company is managed, and these changes must come from the top. For example, the underlying concept of continuous improvement requires that everyone in the company be trained and educated in agile techniques, and that these techniques be introduced into the plant with the full cooperation and involvement of the entire work force. Use of such continuous-improvement techniques as quality circles creates an environment in which everyone can participate in the revitalization. All suggestions and ideas can be brought to bear to solve quality issues, scheduling problems, and customer-satisfaction questions.

These techniques cannot be introduced selectively; it is all or nothing. Quality, just-in-time (JIT), and empowerment of the work force go hand-in-hand. Many companies in the 1970s got the message on quality and instituted quality-improvement programs. Many of these programs failed or were less than successful because they were not accompanied by fundamental changes in the way the company was managed. Many larger companies tried to enforce JIT deliveries from their suppliers and, in many cases, were successful in forcing suppliers to comply. Suppliers frequently increased inventories on items they were required to deliver just-in-time, thus defeating the agile objective in their own companies. These half-hearted attempts do not bring about the kind of improvements required for agile manufacturing.

Velcro Corporation is a supplier to General Motors. The self-gripping material invented by Velcro is used to hold trunk linings and other fabric in place. When GM first began to require zero defects from its suppliers, Velcro along with many other vendors instigated additional quality control on the products being manufactured for General Motors.[1] The product was inspected to more stringent standards, and material that did not reach the standard was discarded. Velcro was happy with its product quality because there were no returns from GM; there were zero defects in GM deliveries.

Velcro managers were shocked when GM criticized their product quality as a part of GM's supplier certification program. GM's criticism was not that Velcro was delivering inferior product, but that Velcro was trying to improve quality by rejecting poor material instead of resolving problems at the source and preventing defects. Discarding material in order to improve quality was not a world-class approach to quality. GM gave Velcro six months to change its production processes to improve quality. During this time Velcro implemented many agile techniques (including Total Quality Control (TQC), quality circles, and JIT manufacturing) and were able to more than meet GM's needs. Agile manufacturing requires fundamental change — not window dressing.

Techniques of Agile Manufacturing

A number of techniques are employed by agile manufacturers to eliminate waste, reduce inventory, and improve cycle times. Those that apply to shop-floor control and production planning are summarized in Table 3-1.

Cellular Manufacturing

For movement of material through the plant, traditional manufacturing relies on a process based on the routing of the product. The shop floor is laid out according to operations performed in a work center. When products are manufactured, production routings determine how products are moved from one work center to the next until the entire production process is completed.

This approach has several disadvantages. First, it requires product to be moved a considerable distance from one work center to the next throughout the process. Movement is wasteful; no value is added while product is being moved. Cycle time is increased because of the movement of materials; typically, there is delay because moving is done by handlers who may not be immediately available. Second, movement creates independent work centers that specialize in one aspect of production (welding, for example) thus limiting flexibility because the people in each work

Agile Manufacturing Techniques Used to Eliminate Waste, Reduce Inventory, and Improve Cycle Time

Waste	Inventory	Cycle Times
Material movement	Reduce work-in-process	Reduce work-in-process
Shop-floor queues	Reduce batch sizes	Reduce batch sizes
Improve quality	Improve quality	Shorten setup times
Processing transactions	Inventory pull	Synchronized manufacturing
Reduce paperwork	Eliminate bottlenecks	Eliminate bottlenecks
Preventive maintenance	Make to order	Preventive maintenance

Table 3-1. Agile Techniques Used to Eliminate Waste, Reduce Inventory, and Improve Cycle Time

center are not used in other work centers. A team approach within the company is also hampered because people feel loyalty to their own work center rather than to the production group as a whole. Third, this approach lends itself to large batch sizes because moving a small quantity of product several times is not economical; it is better to make and move a large quantity once. This process runs counter to agile manufacturing's ideal of reducing batch sizes.

Agile manufacturers usually lay out the shop floor into production "cells."[2,3] A production cell consists of several machines and/or operations arranged together into a cell that is dedicated to the production of one type of product. All machines or operations required to produce a particular family of products are located together in the cell. This approach eliminates material movement because the entire production process occurs within the same area, and materials are simply moved from one machine to the next within the cell as products are manufactured.

Cellular manufacturing also aids production flexibility because people within the cell are trained in several different skills and are able to work on the entire production process for that family of products. The rate of production of that product family can also be flexible because the number of products flowing through the cell can easily be changed. Often production cells are laid out in a U-shape so that materials enter the cell at one side and flow around the U-shape. This shape makes the rate of production easier to change. If high volumes are required, a large number of people can be assigned to the cell; for example, one person for each job step. If the production rate must be reduced, fewer people are assigned to the cell, and each person performs two or three steps. Because the cell is U-shaped, people can easily move from one machine to another as materials flow through the cell. U-shaped cells, in contrast to production lines, have the added advantage that people work together in close proximity, allowing development of a cooperative team spirit. A further advantage of U-shaped or similarly designed cells is that work is visible to other people within the cell and the plant. The production process can be understood easily and problems identified quickly because they are visible.

Cellular manufacturing greatly simplifies production scheduling because work is not scheduled through a large number of work centers, but through a smaller number of cells. Fewer process steps must be planned because the entire product or major subassemblies are manufactured through the cell. There are fewer item numbers because there is no need to identify the parts fabricated within the cell or the subassemblies created within the cell. These items are used immediately in the next stage of manufacture.

Inventory is reduced by cellular manufacturing because the number of parts and subassemblies in inventory is reduced. A traditional manufacturer will raise work orders (or job orders) to manufacture fabricated parts and subassemblies. These items are then inventoried, waiting to be issued to a final assembly work order. With cellular manufacturing the whole product is manufactured within the cell, making it unnecessary to hold parts or subassemblies in inventory.

Point-of-Use Storage

An agile manufacturer endeavoring to reduce inventory and material movements frequently prefers to keep materials on the shop floor instead of in a stockroom. Because there will be very little inventory and much of it delivered just-in-time directly to the shop floor, materials can be kept adjacent to the work cells that use that material. Often there is an area for raw materials and components for each individual cell. These storage areas are called "inbound stock points." All materials used in the cell flow through the inbound stock point. Material can be delivered directly from a vendor to the inbound stock point, a previous cell can deliver subassemblies into the inbound stock point, and material can be moved from a stockroom to an inbound stock point.

Items manufactured in a cell are delivered to the cell's "outbound stock point." If these items are finished products, they can be used to fill customer orders. If they are subassemblies, they can be used to supply an upstream cell. Outbound stock points are often used within a *kanban* approach because the outbound stock point itself can be used to regulate the amount of product manufactured. The number of containers (or kanbans) assigned to a product can be shown in the outbound stock point, and people in the cell will know when to manufacture an item by seeing how many kanbans are open in the outbound stock point. For easy and visual recognition, spaces for kanban containers can be drawn on the floor or on the racking in the outbound stock point. (The kanban method is discussed in detail later in this chapter.)

Software for Cellular Manufacturing

Software systems used by agile manufacturers employing production cells must be able to support cellular manufacturing. Traditional MRP II systems usually allow the definition of work centers, and production jobs are scheduled through work centers. Many systems have the added feature of grouping work centers together into types. This feature can also be used to group work centers together into cells. However, one important characteristic of a cellular manufacturing approach is that production scheduling is simplified. Instead of having to schedule a large number of individual work centers, a small number of cells are more easily and quickly scheduled and rescheduled. Ideal production-planning systems will identify production cells and work centers as separate entities.

Sometimes it is useful to identify individual machines or work centers within a cell and to be able to show these on the production routing. This process is necessary for defining the production process by production engineers and for calculating standard product costs through a cost-buildup calculation. The production cell is used for planning and scheduling on the shop floor; the detailed production routing is used for product definition and cost buildup.

The ideal production-planning system will define production cells and will enable production to be scheduled through those cells. These production cells can be identified with a cell number, a description, and a cell group. Production capacity can be defined by a production calendar, production hours, number of shifts, crew size, and productivity factor. Products manufactured within each cell can be

---

Cellular Manufacturing

The transition from scheduling work orders through work centers to scheduling through cellular manufacturing requires fully integrated software. It must be possible to manufacture the same product in the same plant at the same time using both work centers and production cells. Therefore, MRP, master scheduling, and inventory control must be capable of recognizing production scheduled either way.

When software has this degree of integration, a gradual changeover from traditional manufacturing to cellular manufacturing becomes possible. A good starting point is to determine a product family that lends itself to cellular manufacturing, design a U-shaped cell for the manufacture of these products, adjust the shop floor layout, and begin manufacturing using the cellular approach. This pilot approach can then be extended to other products and product families.

defined with the production rate, cell load, and yield information. Although a product may be manufactured in many cells and a cell may produce many products, a primary cell must be defined for each product so that MRP or master scheduling can automatically schedule products to cells.

The software must be able to identify inbound and outbound stock points as part of the cell definition and have override capabilities for individual products. For example, suppose that all products manufactured in a cell are delivered to one outbound stock point — except one product which for some reason requires a different outbound stock point. Or another product may have specific safety requirements and require separate stocking. The system must be able to accommodate this flexibility.

Companies using stockrooms instead of point-of-use storage can define the stockroom as the inbound stock point. If necessary, the outbound stock point of a cell can be defined as the inbound stock point of another cell (for example, when the output of one cell is used in the manufacture of products in an upstream cell). This makes moving an additional stock transaction unnecessary. The flexibility must be available either to require a stock transaction or to allow direct transfer.

When entering a "production completion" for a cell, the operator reports the cell, product, and quantity. The system then performs at least three transactions: (1) the finished product is moved to the outbound stock point, (2) the components are backflushed from the inbound stock point, and (3) the production schedule is updated.

Distance-Moved and Next-Work-Center Reports

When a company lays out the shop floor into production cells, a clear understanding of the flow of materials through the plant is necessary. One objective of cellular manufacturing is to minimize material movement throughout the production process. In many cases, cell layout is obvious because the company manufactures a limited range of products, and these products fall nicely into families with similar production processes. In other cases, particularly where there is a wide range of products, detailed analysis may be necessary in order to establish the optimum cell configuration.

A "distance-moved report" can analyze the production process to determine the distance materials are moved within the plant. A "next-work-center report" analyzes the movement of materials to determine how often one work center follows another in the production process. While similar, these two reports are sorted in different sequences. Most production-control systems contain information that defines the product and how it is made. The product definition usually consists of a bill of materials (recipe, formulation) and a production routing (or process sheet). The bill of materials defines components or ingredients of the product. Production routing defines how the product is made by identifying which work centers, what processes, and how much time are required in the manufacture of the item. These production routings can be analyzed to determine the distance moved and the next-work-center information.

Calculation of the distance moved necessitates knowing the distance between each work center. Because this information usually is not available in a standard system, a table showing the distance a product must be moved from one work center to the next must be created. A plant with a simple layout can use a grid. The grid reference of two work centers provides enough information for the distance to be calculated. A grid reference is easy to administer and requires considerably less data entry and maintenance than a table showing the distances between each pair of work centers.

Because most factories are not simple enough for a grid approach to be accurate, a distance-moved table containing the distance between all work centers is required. Creating and maintaining this table can be a considerable task if there are a lot of work centers involved. However, once the table is available, analysis reports are easy to create. This table must be maintained accurately if the analysis is to be effective.

Analysis reports can be created from the master production schedule, which contains details of future planned production, or from the history of completed production that shows what has actually been produced in the recent past. Production routings for items being manufactured are interrogated to determine the work centers involved, and distance moved is calculated by referring to the distance-moved table. The report totals the distance moved from one work center to another and provides subtotals by product group or section of the production plant.

A careful study of the distance-moved report will show the most advantageous arrangement of work-center operations to minimize material movements. The next-work-center report provides information about the frequency of movements between work centers. "What-if" analysis can be performed by designing a cell layout, entering the new distances into the table, and running the reports a second time over. In this way, various designs can be analyzed to determine the optimum layout for minimizing the distance materials are moved through production.

Rate-based Scheduling

The introduction of cellular manufacturing modifies a traditional discrete manufacturing process making it more like a repetitive or process manufacturer. For years, process manufacturers have planned and scheduled production according to the production rate through the plant rather than assigning work orders to specific batches. The rates may be hourly, daily, weekly, or even monthly.

Rate-based Scheduling

Matching production to customer demand cannot be achieved overnight. It requires short cycle times, small batch sizes, and flexibility of production. A useful starting point for matching production to customer demand is to base production schedules on recent sales — make today what you sold yesterday. In reality, companies start by scheduling this month what was sold last month (with adjustments for seasonality and changing safety-stock levels). Then they change to scheduling over a two-week, weekly, or half-week period, and finally attain a daily schedule of making today what was sold yesterday.

As the scheduling period is reduced from one month to one day, the amount of adjustment to the schedule falls dramatically. When half-week or daily scheduling is achieved, production scheduling becomes simple. With less emphasis on forecasting the unknown future, more emphasis is placed directly on meeting customer needs.

Agile manufacturers use techniques of rate-based scheduling for planning production for two primary reasons — to level production flow through the plant and to synchronize production with customer demands. The argument goes like this. The final customer (the ultimate consumer) purchases products in fairly even quantities. Products may be affected by seasonality and market fluctuations, but demand for most products does not vary greatly from one day to the next or one week to the next. If a company manufactures a range of products and those products are purchased in an even flow, then the product should be manufactured evenly at the same rate the customers are buying.

Making products in line with real customer demands has many advantages. It synchronizes production, allows for uniform plant loading, and provides a repetitive cycle of events that lends itself to quality and productivity. Repetitive cycles provide for improvement in many areas including marketing and engineering, and an agile objective is to bring various activities of production into synchronization; "to schedule so as to capture maximum possible advantage of regular cycles to improve every aspect of operations."[4]

The outworking of these ideas is to establish production schedules based on customer demand and to make products according to demand patterns of the market. As a simple example, consider a company making three different products. If the average demand is 120 per day for product A, 240 for product B, and 60 for product C, then the company should make 120, 240, and 60 respectively. The economic order quantity (EOQ) for product C may be 1,200; the agile manufacturer would work on the setup times and batch sizes so that product C can be made in lots of 50 or less. The next step would be to reduce batch quantities so that one-third of a day's demand can be made each shift (assuming three shifts) in a cycle of 40:80:20. As batch sizes drop further, the production pattern can become even more streamlined. The ultimate goal would be to manufacture products A, B, and C respectively in a cycle of 2:4:1 — repeated again and again.

A company that has reduced batch sizes to very low levels can achieve a uniform load on production throughout the production process. In the previous example with the final-assembly pattern of 2:4:1 repeated 60 times per day, the load on upstream production cells (e.g., fabrication, sub-assembly manufacture) will be entirely uniform and predictable. Demands on vendors will also be highly predictable throughout the day. This level of uniformity produces, in and of itself, high quality and continuous improvement. The process is so much under control that people are able to apply themselves to improvement.

Of course, this is an idealized example. Nevertheless, the principle holds true that agile manufacturers place great emphasis on small batches, uniform flow of production and plant loading, and continuous improvement. Another aspect is that the process is so much under control that production is in tune with changing customer needs. As the total demand varies, the production rate can be increased or reduced; as the mix changes, elements of the production pattern can be adjusted.

As production takes on the characteristics of repetitive flow (even though the products may be far from repetitive products), creating work orders for each batch becomes wasteful and unnecessary, especially as batch sizes are reduced to very small quantities. Production must then be planned using rate-based schedules. These schedules show the quantity of a product to be manufactured during the day or the shift. They are the plans that drive MRP to supply sub-assemblies, components, and raw materials. They are the schedules that are updated when production completions are reported, and they can be the vehicle for informing operations personnel of today's production plan.

Software for Rate-based Scheduling

Software needed to support rate-based scheduling requires the ability to create rate-based schedules, make changes to individual schedules as a whole, and provide inquiries and reports. These schedules need to be fully integrated with master production scheduling (MPS) and materials requirements planning (MRP) and to allow mixed-mode manufacturing with work orders and rate-based schedules working side-by-side.

Rate-based schedules can be created by the MPS application, the MRP application, or manually. In reality, rate-based schedules for final assemblies are created manually by the production planner based on customer needs. MPS and MRP can then be used to "explode" final assembly schedule requirements back to upstream production cells or work centers. Because cellular manufacturing has relatively few cells in comparison to traditional work centers, the number of production areas that need to be scheduled is reduced; therefore, creating the final assembly schedule using rate-based schedules is quite practical. However, the software must have programs that enable schedules to be created and changed quickly and easily.

The program to create rate-based schedules will require entry of the product, the cell, and the rate information. Having several different methods to create schedules according to plant needs is helpful. The simplest method is for the planner to enter, one day at a time, the rate required for that day. This method is time-consuming but does allow for the rate to be different each day. Another method is to enter the date production should start, the date production should end, and the total quantity to be made. The system will then calculate the average quantity per day and set the schedule for each workday equal to the average. The fixed-rate method allows the entry of the fixed production rate for the product and the start/stop dates. The system sets that fixed rate for each workday within the dates. The fixed rate will be set as a default on the item master for each product. The forward and backward schedule methods rely on entering the start date or the stop date, the total quantity, and the fixed daily rate. The system then calculates the number of days to manufacture the quantity and sets the fixed rate in each day, either forward from today or backward from the stop date.

Rate-based schedules may be created by master production scheduling from customer orders or forecasts. Instead of creating suggested work orders, MPS creates suggested rate-based schedules. An indicator is required on the item master record to show if a product is manufactured principally with work orders or rate-based schedules. MPS uses this indicator to determine what kind of production plan to suggest. Similarly, when an item can be made using more than one cell, a primary cell must be assigned to show MPS where to create the suggested schedule.

It can be useful for the MPS recommendation to exactly match customer orders in accordance with the usual rules of scheduling (including rounding to batch sizes), minimums, pack quantities, kanbans, and so forth. However, it can also be useful for MPS to automatically spread or level the recommended production schedules. When traditional rules are used, MPS recommendations will not be evenly spread across the days because customer orders will not be evenly spread. The scheduler must then adjust the recommendations manually to achieve a level schedule. When automatic leveling is employed, the MPS calculations will attempt to spread uneven recommendations across the time period according to predetermined rules. These rules may include using standard rates for each product, averaging the quantities for each product, or using forward and backward scheduling using fixed rates. In reality, devising satisfactory methods of automatic spreading without resorting to finite scheduling or simulation techniques (which are discussed later) is difficult. Nonetheless, simple averaging provides a useful starting point upon which the production scheduler can build.

Schedule Status

Just like work orders, having different statuses for different kinds of rate-based schedules is helpful. Minimum requirements are: suggested schedule, firm-planned schedule, and released schedule. Suggested schedules are those schedules created by MRP or MPS as suggestions to the production planner. These are converted to firm-planned schedules after the planner reviews requirements and recommendations. A firm-planned schedule has been approved by the production planner and authorized for the procurement of components and production capacity. Reports, inquiries, and available-to-promise calculations include firm-planned schedules because they represent planned future production. A released schedule is one that has been released for manufacture on the shop floor, authorizing shop-floor personnel to make the item. Additional useful statuses may include a "what-if" status used by the planner for simulation purposes, an "on-hold" status for schedules frozen for one reason or another, and a "pending-deletion" status.

Changing Rate-based Schedules

Setting up rate-based schedules, although quick and simple, does create a large amount of data to be held on a production-schedule file. Instead of a single work order covering several days of production, there is a scheduled quantity for each day or shift. There needs to be a method to easily manipulate this information so that significant changes can be made with a minimum of manual data entry.

This function for changing rate-based schedules must be able to select ranges of schedules for a cell, a group of cells, a product, and/or specific statuses. The program must allow these schedules to be manipulated by changing their status, moving the schedule dates in or out, and increasing or decreasing the quantity.

The primary use for changing the status of a series of schedules is to release a day's worth of schedules by converting them from firm-planned to released schedules, for example. Moving schedule dates is used when customer requirements change, or to accommodate problems like machine breakdown when the entire production needs to be moved from a cell for one day.

For example, the quantity increase/decrease feature, which allows for scheduled quantities to be increased or decreased by a set amount or by a percentage, is used when specific changes occur in customer requirements or possibly when an item on promotion has a much bigger (or smaller) response in the marketplace. This change is achieved by selecting the range of schedules to be changed by cell, date range, product, planner code, and/or status. The changes are then given either a quantity or percentage, and the system will make the change to all schedules that fall within the selection criteria.

Materials Requirements Planning

After the master production schedule has been created successfully by the production planner, materials requirements planning determines which subassemblies, fabricated parts, and raw materials are required to complete the master schedule for each product. MRP takes rate-based schedules for the master-scheduled items and explodes them down to their constituent components, subassemblies, and raw materials. MRP calculates quantities and required dates for each lower-level item by exploding through various levels of the bill of materials and by using lead times from the item master file for each part or item. These calculations create suggested orders for these lower-level items. The suggested orders will include rate-based schedules for items flagged as being planned using schedules, work orders for items flagged as such, and purchase requisitions for items bought from outside vendors.

If the final assembly schedule has been planned carefully with a level loading on the final assembly cells, then the explosion through MRP will create level or uniform loading on the upstream cells and work centers. This explosion of a uniform set of schedules enables upstream work cells to be synchronized to the final-assembly production which, ideally, has been synchronized to customer needs. For this reason it is important and desirable for the planners to spend their time ensuring a well-synchronized, levelized schedule for final assembly. When this levelized schedule is "blown through" MRP, the entire plant and the vendor deliveries are synchronized with final assembly.

The production planning and control system must handle the combination of work orders and rate-based schedules. When agile techniques are being introduced, it is often impossible to immediately eliminate work orders and convert to the exclusive use of rate-based schedules. It is helpful when MPS, MRP, and system reports and inquiries incorporate supply and demand from both work orders and schedules. In addition, moving all production onto rate-based schedules is sometimes not possible or desirable. For example, final assemblies might be controlled with work orders and the lower-level assemblies manufactured using rate-based schedules, as when detailed tracking is required for final assembly, perhaps owing to military or Food and Drug Administration (FDA) regulation or to a design-to-order process using standard parts and subassemblies. MRP and master scheduling must be able to accommodate this kind of mixed-mode manufacturing.

Capacity Planning and Cellular Manufacturing

Capacity planning and detailed production scheduling are simplified by agile techniques. Production capacity and capability is more flexible because cycle times are shorter, batch sizes are significantly reduced, and operators are cross-trained to allow for last-minute mix and volume changes. In addition, a working environment that values customer needs can be established to the point that the people are motivated and willing to do what it takes to meet those needs. This environment frequently requires an organization in which the production operators and supervisors have full responsibility for meeting production schedules. Some companies have policies that no one goes home until every customer order is shipped; conversely, when orders are light, the plant closes and the people go home irrespective of how many hours they have worked. This level of trust, commitment, and team identification — while rare and taking years and significant leadership to engender — is a goal of agile manufacturing.

The two levels of capacity planning used within any production plant, whether traditional or just-in-time, are "long-term capacity planning" and "short-term capacity planning." Long-term planning consists of forecasting sales over a six-month, twelve-month, three-year, or five-year period and determining what capacity is needed to meet customer demand. Short-term planning is the detailed planning and loading of production onto the work centers or cells within the production plant. Both agile and traditional manufacturers approach long-term capacity planning similarly — through the master-production-scheduling process, often using rough-cut capacity-planning analysis to match demand to capacity.

Rough-Cut Capacity Planning

In rough-cut capacity planning, the computer system calculates the resources needed to manufacture all the products in the forecast and matches that to the resources available with the plant. The rough-cut aspect of this analysis is that the system does not calculate every detailed step in the manufacturing process; only certain resources or products are considered. For example, rough-cut capacity planning may be performed at the product-group level. Instead of forecasting and analyzing each individual product, these products are consolidated into product families for forecasting and capacity planning purposes.

Capacity Planning

Capacity planning is detailed and complex for traditional manufacturing plants and leads to considerable waste and confusion. There is constant rescheduling and expediting, queue buildup, and missed delivery dates.

Agile manufacturers find capacity planning less burdensome because short cycle times, production flexibility, process simplification, and committed people eliminate most problems. Apply WCM methods in the plant and capacity almost takes care of itself.

If, for example, cola drinks are sold in two-litre, one-litre, and eight-ounce containers, then, for rough-cut capacity planning purposes, analyzing each product separately is not necessary. A forecast and resource analysis at the product-family (cola) level is usually adequate. Similarly, if production has plenty of capacity in all areas except for heat treatment, then the heat-treatment ovens are the bottleneck operation within the plant. For capacity-planning purposes, analyzing products based only on their use in the bottleneck operations may be enough. The limiting resource (such as a particular machine, a scarce raw material, or available cash) is the only relevant factor to examine.

Detailed Capacity Planning

A traditional manufacturer will approach detailed capacity planning by having a system that studies the total manufacturing lead time of the products being made and analyzes the process work order by work order, work center by work center. Printed reports compare the load and capacity on each work center. The planner's job is then to reschedule the work orders to match available capacity — a difficult, often impossible task because there are hundreds of products and sub-assemblies being processed through a large number of work centers. Because each schedule change causes further reper-cussions, change on other work centers can only be fully accessed by running the capacity planning a second time over.

Many difficulties experienced by traditional manufac-turers in capacity planning are caused by long lead times and complex production process. Cellular manufacturing allows capacity planning to be handled quite differently. Because there are far fewer production cells than traditional work cen-ters and shop-floor layouts are simpler, the capacity calcula-tions are fewer and simpler. Capacity-planning programs do not have to schedule production through work centers (the way a traditional MRP II system requires) because scheduling is done manually or directly using MRP or master schedul-ing.

Even more helpful is an on-line capacity-planning program that identifies capacity issues as they arise. It is help-ful for the system to show the capacity being used in the cell and to show where capacity is available during the schedul-ing process. Thus, when manually scheduling the cell, a plan-ner can schedule a cell according to available capacity.

Table 3-2 shows one example of an inquiry that can be used for this purpose. The planner enters a product and a cell. The program calculates the capacity on the cell each day, the cell loading (for all products scheduled), and the available capacity. Available capacity is then converted into the quanti-ty of entered product that can be manufactured during that day. Capacity, load, and available quantity are shown cumu-latively for each day and averaged across the days. The plan-ner can use the cumulative and average information to determine when the product can be scheduled on the cell.

```
Cell (F331)      Final assembly line 6        Dept 300      Plant (Brom)
Product (1012) square hose reel
     Statuses ( ) ( ) ( ) ( ) ( ) ( )    blank = ALL    Thru date (021593)
     What-If  %load  change utilization   crew size
```

	1/23/93	1/24/93	1/25/93	1/26/93	1/27/93
	MONDAY	TUESDAY	WEDNESDAY	THURSDAY	FRIDAY
PER CAP	86.40	86.40	86.40	86.40	86.40
PER LOAD	53.12	92.50	3.12	0.00	0.00
PER AVAIL	33.28	-6.10	83.28	86.40	86.40
PER QTY	15.00	-2.00	37.00	39.00	39.00
CUM CAP	86.40	172.80	259.20	345.60	432.00
CUM LOAD	53.12	145.62	148.74	148.74	148.74
CUM AVAIL	33.28	27.18	110.46	196.86	283.26
CUM QTY	15.00	13.00	50.00	89.00	128.00
AVG CAP	86.40	86.40	86.40	86.40	86.40
AVG LOAD	53.12	72.81	49.58	37.19	29.75
AVG AVAIL	33.28	13.59	36.82	49.22	56.65
AVG QTY	15.00	6.00	16.00	22.00	25.00

Table 3-2. Available-Capacity Inquiry for a Production Cell

In the example shown in Table 3-2, if the planner needs a quantity of 80 for product 1012, this amount can be scheduled for completion on Thursday because that is the day the cumulative available quantity exceeds 80 and the capacity is available to complete the quantity by that day. The inquiry shows when the product can be made and how to schedule the cell to make the product. If this cell lacks enough capacity, the scheduler can review other cells capable of making the product to determine their available capacity.

The flexibility that is often available in an agile work environment allows the scheduler considerable leeway when loading cells. The inquiry has some "what-if" capability that enables the scheduler to easily test out changes to crew size, utilization percentages, and load. One big advantage of cellular manufacturing is that the production rate can often be adjusted quite easily; more people can be added to a cell, or the machines themselves can be sped up or slowed down. This flexibility of production rate enables operators to synchronize manufacturing output to meet customer needs.

Finite Scheduling

Many attempts have been made to develop automated techniques to replace the skill and knowledge of the production scheduler. MRP II can do most of the shop-floor scheduling in a traditional manufacturing environment, but most of these systems employ infinite-capacity-planning methods. Infinite planning does not attempt to match demand on the plant to the available capacity at a cell, work center, or even total plant level. It assumes that all orders entered can be made using available resources. This assumption is often good, if the company has done the longer-term rough-cut capacity planning, and if customer orders do not significantly exceed forecasts used for that planning. This approach, however, always breaks down when it comes to the details. Individual work centers become heavily overloaded while other work centers are under-utilized. Queues develop on the shop floor, lead times become elongated, shipment dates are missed, and customers become angry. This dilemma gives rise to the traditional habit of expediting orders, which is time-consuming and wreaks havoc with any semblance of production planning. Frequently, products are rushed out at the last minute (or at the end of the financial period) and often quality is compromised in the name of customer service.

At first it would appear that such problems could be overcome by a sophisticated computer system that not only analyzes when products and subassemblies are needed, but also keeps track of available capacity within work centers and schedules orders so as not to overload any work center. These kinds of computer systems are available and have been used successfully in some companies. Often they are not successful because the technique has some basic limitations, including the assumption that work-center capacity can be precisely defined and that the load put on a work center by the manufacture of a product is well understood.

In reality, the capacity available within a work center is not well defined. The number of hours in a day does not define available productive time. Skills of individuals vary considerably in terms of speed and productivity. Machine productive capacity changes over time and in accordance to the quality of preventive maintenance. Quality of products produced and quality of components going into those products significantly affect the "real" capacity of the work center. In addition, the definition of working capacity cannot adequately take account of people's willingness to work extra time or work faster when the work center is under pressure.

Similar problems occur with defining the production-load created by the manufacture of a batch of a specific product. Most manufacturers, particularly traditional manufacturers, lack well-developed, complete, and accurate bills of materials and production-routing information. The information is usually determined and used by either the design engineers or the cost-accounting people who are not concerned enough about details of production for the needs of finite scheduling. Even when bills and routings are meticulously maintained by production engineering and scheduling personnel, keeping this information up-to-date and accurate is very difficult.

For finite scheduling to be successful, this information must be accurate. Finite scheduling has been most successful within production plants where accurate information is easier to provide. These plants may have highly structured process manufacturing where capacity is well defined and where product characteristics are clearly understood.

In general, finite scheduling has not provided the startling benefits envisioned by its pioneers in the 1970s. However, advances are currently being made in this area and many new products are becoming available that combine simulation, scheduling, and artificial intelligence (AI). The AI aspect of these systems can take the form of the system "learning" from what actually occurs and gradually, over time, making better predictions and suggestions because it accrues more and more information about true production capacities and loads. Alternatively, the artificial intelligence can be an "expert" system.

An expert system relies on human experts (in this case, schedulers and planners) to provide detailed information so that it can begin to learn how to make suggestions that reflect the thoughts and experience of the experts. Similar systems are also used in the medical profession for computer diagnosis. The system knows nothing about medical issues until a physician provides detailed information about the nature of ailments and diseases. As more and more information is provided to the system, the diagnoses become more meaningful. In a similar way, production schedulers gradually teach the system all the details and nuances of scheduling so that the system can make useful suggestions. These simulation and finite-scheduling systems are still being developed, and while there are a number of successful implementations, they have yet to prove their value in a wide range of production situations.

Finite Scheduling in Cellular Manufacturing

The limitations of finite scheduling are the same in agile and traditional manufacturing environments. However, finite scheduling can be adapted more successfully to cellular manufacturing because some limitations of the technique are alleviated. Agile manufacturing frequently has much better control of the production process and, through the mechanism of continuous improvement, has more precise information and understanding of how the product is manufactured. As production cycle times are shorter, the period being scheduled is generally shorter. Scheduling is also less complex because schedulers are dealing with a few cells instead of many work centers. The precision needed for finite scheduling is easier to achieve when dealing with short time periods and fewer locations.

On-line finite scheduling of an individual cell can prove most successful. The master schedule or MRP systems automatically load work cells with suggested orders. On-line finite scheduling can be used to spread the suggested orders over the scheduling time period. The mechanism used to spread the orders varies according to the situation, and programs to do this spreading need some flexibility to accommodate different algorithms for use under different circumstances.

Options for Finite Scheduling

There are three options for establishing an approach to finite scheduling of orders. The first choice is governed by the extent to which schedules must adhere to original production dates. If orders are being spread across one week and schedules are already set for certain days, the system should give the choice of either spreading the schedules over the entire week or of attempting to keep orders within the originally set days. If scheduling is to be done over the entire week, then the original dates will be abandoned and schedules reassigned according to other priorities and the resource constraints. If daily dates must be retained, then the system will attempt to prioritize orders within each day and will report any overload.

The second choice involves "frozen" orders. The scheduler may be planning to reschedule the entire cell over the following week, but there may be some schedules on the cell that should not be altered. These decisions typically will be made based upon the status of the orders. For example, the scheduler may decide that all released schedules should be frozen and that only firm-planned and suggested orders be rescheduled. There may also be specific orders that the scheduler wishes to freeze. The system will then schedule around these frozen schedules.

The third choice determines whether the order should be averaged across the week, forward scheduled, or backward scheduled. If orders are averaged across the week, then an even amount of work will be assigned to the cell for each working day within the week. If they are forward scheduled, then the first day of the week will be fully loaded, the second day, the third day, and so on until all production has been scheduled. Backward scheduling works the same way except the last day is loaded first and the process works backward toward the first day of the week until all production is scheduled. If there is more work scheduled than there are theoretical resources, then the system will warn the scheduler and give the choice of either spreading the excess load evenly over the whole week or back-loading it into the last day. Of course, the scheduler will have to resolve the issue either by increasing the capacity through overtime or by increasing the production rate on the cell, or by assigning the production load to another cell.

If the decision is made to spread the load evenly over the week, then further choices are required. The spread might be intended to evenly load the week by putting highest-priority jobs first. Another option is to have an amount of each product scheduled for each day. This method fits nicely into the concept of level loading and uniform scheduling: Make some of every product every day. The scheduler will add up all products being made, calculate the amount of capacity required to make these products, and then prorate available capacity on the first date in the same proportion as the total amount of load for the time period. This proration is repeated each day until the entire amount is assigned.

Once the approaches that can be taken for finite-order scheduling have been established, the method of prioritization must be addressed. The method will vary significantly by company and by plant. One method is to base the priority on the original required (or requested) order date. Another method is to assign different priorities to different customers and/or different products. In a make-to-stock environment, a simple method of priority setting is the amount of available stock of the product; the lower the available stock, the higher the production priority. There are other specific priority rules pertaining to the company or the products; for example, the physical size of the product, delivery-route schedules, or industry traditions.

Major and Minor Setup Times

A very important aspect of sequencing is the setup times required between products. Agile manufacturers place considerable importance on setup time reduction to ensure the flexibility needed to quickly and effectively move from one product to another and to manufacture small batch sizes in order to retain low inventories. When this flexibility has been achieved, setup time problems become less significant; however, there may still be issues relating to major setups.

Typically, a major setup occurs when the cell changes from making one kind of product to another product. This changeover may require retooling, rejigging, adjustment of machines, and so on. A minor setup, such as a change from one product to another within the same family, usually requires much less time and effort. Often it is important to establish a schedule that minimizes major setups. In other words, similar products (that is, those requiring minor setups between them) must be scheduled together. This kind of scheduling issue can be further complicated by the need to schedule products in a particular sequence within the family, as, for example, when products are manufactured in a range of colors and the schedule needs to be set up so that colors go from light colors to dark colors throughout the scheduling cycle. This approach minimizes minor setups because fewer wash downs are required.

Having standard scheduling software that can account for all these specific scheduling rules is not possible. However, being able to establish within the software the appropriate fields that are then filled with the required information is possible. Custom programs can be written filling these fields with the company-specific data (for instance, light to dark codes, days of stock, major/minor product families), and the standard system merely sorts and sequences according to the content of these fields. This system provides the best of both worlds. Standard software is used, but the information is provided taking account of the unique characteristics required by the company and the products.

Finite Scheduling Over Many Cells

There are limits to the usefulness of finite scheduling in a multicell product plant because of the limitations of the information available to the system and because of the complexity of the problem. These limitations can be approached by creating an increasingly complex scheduling model and attempting to take all criteria into account. This approach is unsuitable for an agile environment because it runs counter to the ideas of simplicity and transparency. A better approach is to provide a finite-scheduling method that is simple and straightforward, takes account of the basic factors involved in the scheduling process, and provides a guide for the scheduler. Finite scheduling of this sort does not attempt to provide a complete solution to the problem, but does provide a framework from which the scheduler can apply his or her judgment and business knowledge.

Multicell finite scheduling is run as a large batch job; it cannot be provided on a real-time, on-line basis because it requires a large amount of calculation and takes considerable time to run. The techniques required are the same as those needed for a single cell, and the same choices of approach must be made. In addition, products should be grouped into product families for scheduling purposes. A field on the item master file is required that specifies this family, and the field must be used for this purpose only. Product groupings used for pricing, sales analysis, or financial consolidations are unlikely to be suitable for production scheduling. Product families are assigned according to major and minor setups.

Items with the same scheduling family have minor setups when moving between products within the family. A major setup is required when moving from one family to another. Finite scheduling will attempt to automatically move products from one cell to another in an attempt to balance the load and ensure that all products are manufactured on time. But finite scheduling will also attempt to keep families together on one cell.

All products have a primary cell that is used by MPS and MRP to create suggested orders. For finite-scheduling purposes a secondary and tertiary cell is required. These cells may be assigned at the family level or the individual product level, and the logic of the finite scheduling will differ according to the level of cell assignment. A table is established so that the finite-scheduling program knows where to reassign product families when a cell becomes overloaded. The logic of the finite-scheduling program is that all schedules on all cells are read by the system; capacity and loads are calculated and overload conditions are identified. The system then reassigns whole product families to secondary or tertiary cells in an attempt to balance and level the load on each cell.

These calculations can be iterative; that is, the system will read through the information and do its calculations and adjustments. It will then read through again to determine the situation after those changes have been made. As a result of this second read, further reassignments will be made to refine the scheduling. This iterative process is repeated several times until an optimum result is produced. The number of iterations required can be set as a parameter of the program, and there are a series of dampening factors which ensures that only significant changes are made. An example of a dampening factor is an overload percentage. If a cell is 10 percent overloaded and the overload percentage is 15 percent, then the finite scheduling will not attempt to change the cell because the overload is less than 15 percent. These dampening factors are important because they prevent the process from becoming unstable and from producing confusing results.

Parameters and dampening factors can be set differently for each iteration. By using a parameter table that allows parameters and factors to be set differently for each iteration, the first iteration can, for example, allow many changes to schedules, the second iteration can be less sensitive, and so on. One of the primary parameters sets the maximum number of iterations. Running the scheduling with a limited number of iterations, perhaps two or three, gives useful results without elongating the run time of the system. Other parameters include the amount of capacity to include in the calculations. For example, iterations 1 and 2 may assume standard working hours each day. Iteration 3 may allow additional hours to be accounted for. In this way, the finite-scheduling program will try to accommodate the whole schedule in two iterations but, if this fails to satisfy the requirements, the third iteration then assumes additional overtime will be made available to meet orders loaded on the plant.

Simulation

A growing amount of software used to simulate shop-floor conditions and to present recommended load-optimization plans is becoming available. These simulation approaches are not new; they have been available for many years. However, the availability of powerful computers and work stations that have a graphical presentation of results makes them more practical than before. Methods used to develop these simulations are not specific to shop-floor control. They have been developed to address all manner of simulation processes from traffic flow on the New Jersey turnpike to battlefield tactical decision making. They are essentially complex mathematical modeling and programming languages that have been given "user friendly" graphic interfaces.

These simulation systems take scheduling information and simulate production flow through the plant from one cell to another and show graphically (or otherwise) the results of the current schedule. The production scheduler can see where problems are likely to occur and can make schedule adjustments. After simulating the adjusted schedule over several days, the results are presented to the scheduler, who makes further changes. This process is repeated until the scheduler is content with the plan.

Many of these simulation systems have finite scheduling built into them. They not only show the results by simulation, but also reschedule cells and present recommended schedules optimized according to rules built into the simulation package. Some of the more sophisticated simulation packages have a degree of artificial intelligence built into them so that the system recognizes the consequences of its recommendations and learns either from the outcome of its actions or from the scheduler. The information the system learns is then used to make better decisions next time the system is used. These "expert systems" and artificial intelligence systems are still in the exploratory stage, but may well become useful and accepted methods of scheduling in the future. The principal problem is that they are complex, require considerable skill and knowledge to use, and are not integrated with other production planning and control systems. These shortcomings may be resolved over the next few years.

Summary

The following issues are at the heart of agile production-planning processes, and computer systems must contain features that assist in the transition from traditional manufacturing methods to the implementation of new agile approaches.

- *Cellular Manufacturing*
  The shop floor is reorganized into production cells that are used to produce a product or family of products. Software must be able to define production cells and account for point-of-use storage like inbound and outbound stockpoints for each cell.

- *Rate-based Scheduling*
  The shop-floor schedule is based upon the amount of production required on a weekly, daily, or hourly basis. Software needs to provide rate-based scheduling features and to fully integrate production planning using work orders and rate-based schedules.

- *Capacity Planning*
  Capacity planning within a cellular manufacturing environment is easier than in a traditional work-center style of shop-floor layout. On-line, real-time inquiries showing the capacity and load on a cell are helpful, and finite scheduling (both single cell and multicells) is a valuable technique that works better with cellular manufacturing.

CHAPTER FOUR

Backflushing and Inventory Pull

C ellular manufacturing and rate-based scheduling are used in agile manufacturing to plan and control production. Inventory pull and backflushing are methods commonly used in cellular manufacturing plants. Both techniques are underpinned by a fundamental change of approach to the control of the shop floor or production facility. A traditional manufacturer is concerned to report every transaction that occurs on the shop floor — material movements, inventory, completions of job steps, and so forth — and feels that the process is under control when there is a detailed flow of information. An agile manufacturer wants to limit the reported transactions because they are not value-added functions; the control required does not come from constant checking, but from short predictable cycles and a clear understanding of the process.

A traditional work-order approach to manufacturing planning and control is designed to "push" materials onto the shop floor in accordance with the preplanned schedule. Most world-class manufacturers approach material movement on the shop floor by "pulling" inventory through the plant as it is needed. The pull method facilitates a just-in-time (JIT) production process.

Backflushing

Agile manufacturers use backflushing for two reasons: (1) it reduces transactions into the inventory control and planning system and (2) it improves the accuracy of information. Backflushing is a technique used for reporting production completions and updating inventory records. The backflushing process performs four tasks through a single simple entry. Completed products are booked into inventory (typically finished goods), the production schedule is updated, components and raw materials used to make the product are booked out of inventory, and lot or serial-number information is created.

In its simplest form, backflushing is done by entering the product manufactured, the cell it was manufactured on, and the quantity completed. The backflushing program validates the product, creates an inventory transaction to show that the product has been completed, and "moves" the product into the next stocking location. If the product is a finished item this stocking location will be a finished-product area or stockroom; if the product is a subassembly or partially completed item, it will be reported into the outbound stockpoint of the production cell. The outbound stockpoint may be defined as the inbound stockpoint of the following cell. Additional information like the date, accounting period, and reference number are often shown on backflushing entry screens and may be required data.

At the same time that the program records the completion, the program reviews the production schedule and reduces it by the completed quantity. If no schedule exists or if the schedule is on another cell, this information is displayed so that any required changes can be made. The backflushing program should allow all legitimate entry of information and give warnings when unlikely information has been reported. However, it should not prevent the entry of legitimate data. On some occasions flexibility requires last-minute manual changes to be made to the schedule, and other times it is not necessary to have production schedules set up at all. The validations should highlight possible errors without restricting the use of the system.

Backflushing Components

Components, raw materials, and subassemblies that have been used to manufacture the product are updated by exploding the bill of materials (BOM), formulation, or recipe. The bill of materials defines the components of the item being manufactured, and the quantities of components to be removed from inventory can be calculated by multiplying the component quantity by the manufactured quantity of the product. An inventory record is created for each component showing the movement of inventory from the inbound stockpoint or component warehouse.

Calculating component inventories is often more complex than simple multiplication. A bill of materials is usually more than one level deep, showing the components going into subassemblies that go into the final product. In reality, even quite simple products may have bills of materials that are as deep as five, ten, or even fifteen levels. Backflushing calculations need to know how far down the bill of materials to explode the components. Most agile manufacturers attempt to "flatten" their bills of material, to remove unneeded levels so that the product structure is simplified. This process of flattening bills is done at the time cellular manufacturing is introduced. At that point manufacturing

---

Backflushing

Although backflushing works best without work orders, a first step can be to use backflushing against a work order. An ideal system allows backflushing against work orders, firm planned orders, rate-based schedules, or no orders at all. This method gives the flexibility required for the transition from work order-based scheduling to a rate-based or orderless approach.

It is also useful if the work order and firm-planned-order backflushing programs automatically select the most recent order and then backflush against that order. This feature simplifies the process because the operator can enter only the product and quantity — not the order number.

---

subassemblies and fabricated components are no longer necessary because the entire production process is completed within one cell and subassemblies are no longer stocked and reported. A completely flat, single-level bill can be easily backflushed.

There are often good reasons for having several levels in the bill of materials. Reading and understanding a BOM that has a logical structure is easier, and frequently engineers need them well structured so that the relationship between product design and the production process can be readily seen. When multilevel bills are being used, the backflushing explosion must know which components and subassemblies need updating. This problem can be overcome in a number of ways. One way is to link components and raw materials to a specific step in the production routing and have the backflushing program explode only specified steps. Another is to have a flag on the bill of material to show if the item is a backflushed item for that product and cell. Both of these approaches are in common use but tend to be complicated and easily out-dated.

A third approach is the use of "phantom assemblies." A phantom assembly is an item in the system that is defined for planning-and-inventory-control purposes only, or to simplify the bill structure; it is never manufactured itself because it is merely a step in the production process. These phantom assemblies can be set up in the bill of materials just like subassemblies, but the backflushing program "blows through" them to their components. Thus a double benefit is achieved because a multilevel structure can still exist within the bills of materials and yet a "flat" bill is seen for production purposes.

Control by Lot Number and Serial Number

Backflushing is simplest when no items require the increased accuracy of lot- or serial-number control because the only information needed to do the backflush is the product, the quantity, and the cell. When lot and serial numbers are involved, additional information must be provided.

If the finished product is a lot-controlled item, the backflushing program (1) will identify a lot/serial code on the item master file and (2) will require the entry of a lot number for the product being manufactured. This lot information will be recorded as a part of the inventory transaction reporting the completion of the product. If the finished product is serial number-controlled, then the same information is required (along with a unique number for each one of the products completed). The backflush program will either require the products to be reported one at a time with the serial information or, if the serial numbers are sequential, will automatically assign serial numbers to the products. If automatic serial numbering is being used, the program will usually prompt the user for a starting number, the "static" information, and the increment to be used.

The use of serial- or lot-controlled components is more complex because the system does not know which lot numbers of a component have been used in the manufacture of the product. Some systems assume that lot numbers are used in a logical sequence (the oldest receipt date first, for example) and backflush using this sequence. This method is only acceptable when there are strict disciplines in place within the production plant to ensure that assumptions are always carried out in practice. The purpose of lot and serial numbering is to provide additional control and information about individual products or batches of products. This objective is defeated if automatic assumptions are built into the reporting process.

Other backflushing programs require the user to enter lot numbers and serial numbers at the time the completion is reported. This approach works well in many situations. The BOM explosion identifies all items to be booked out of inventory and determines which of these items are lot or serial controlled. The non-lot/serial items are backflushed in the usual way, and the system prompts the user for the lot numbers or serial numbers of the components.

This approach has some limitations because it assumes that the people entering the completion information know the lot and serial numbers involved. Of course, this assumption is valid in many cases, and in some industries considerable emphasis is placed on the responsibility of the production-cell personnel to take care concerning which lot numbers are used. The pharmaceutical industry is a good example because health-and-safety regulations and FDA requirements are strict, and adherence is a significant part of the production procedures. In other industries, cell personnel need not be concerned about lot numbers, and serial numbers and a two-step backflushing process can be used. The two-step process allows all items to be backflushed irrespective of their being lot or serial controlled. Lot- and serial-number information is entered at a later stage. The system keeps track of lot/serial items that have been issued but have not yet been assigned numbers. Reports and inquiries are available showing that items have been issued, but the lot/serial information is not yet available.

The Use of Lot and Serial Numbering

It would be convenient if lot and serial numbering were unnecessary. Indeed, an objective of agile manufacturing is to simplify production in all its aspects. This simplification would include not using lot and serial numbers if they do not serve a valuable purpose. However, lot and serial numbering is becoming an increasing requirement for finished products because they provide a valuable method of tracing the source of products that are defective or cause other problems. Many regulatory agencies (including the Food and Drug Administration, the aerospace and defense industries, and increasingly the automotive industry) are requiring more stringent traceability information.

At its most basic level, an agile inventory-control system must be able to track lot and serial numbers of items from receipt of raw material, through manufacture, to delivery of product to the customer. Raw materials and components received from vendors often have the vendor's lot numbers on them; these numbers can be used to track those items within the plant. Alternatively, lot numbers may be assigned at the time of receipt, and information like the date, time of receipt, and vendor number commonly is built into the lot number. This receiving lot number is then tracked throughout the system as the material is moved, issued to production, backflushed, or sold.

When products are manufactured, lot and serial numbers can be assigned to them as subassemblies or as finished products. These lot and serial numbers are then tracked as the product is moved, picked, and dispatched. The inventory-control system identifies that the item is lot or serial controlled and requires lot/serial numbers to be entered with any transaction.

A further extension of lot and serial numbering is to have a lot/serial "traceability" system. Here the idea of numbering is taken a step further by building a product structure for each serial number or lot number of a manufactured product. This structure is similar to a bill of material except that it records the components and raw materials used to make a product together with the lot and serial numbers of those components. This information is recorded any time a product is reported complete or when component and raw material issues are made to manufacture a product. A structure of this kind provides complete traceability of what went into each product, where that component came from, and which customers received a specific lot number of a product.

When there is any question about a product, the system can be interrogated to determine which customers received the product and what the constituents of that product were. Other products containing the same lot numbers of components can be traced. This procedure can be used, for example, when a fault is discovered in critical items like automobile brakes or where a food product may have been contaminated.

The production structure (or tree) required for this level of lot/serial traceability can also be used to track other information relating to the product. For example, quality-control data can be entered at each level within the structure along with other relevant free-form text. This feature is useful in regulated industries where specific test criteria are required, or for gathering information to verify the quality certification required by a customer. Because specific information recorded will vary considerably from one industry to the next, computer systems should allow for individual definition of information to be collected and should be able to make the information required or optional depending upon the circumstances.

The production-tree information can be used to record all components used to manufacture an item, irrespective of whether they were lot/serial-tracked components. This information is useful when products vary from one production run to another and the company wants to keep precise track of the content of a product or batch of products. The lot/serial tracking system will have a switch enabling users to track lot numbers, serial numbers, and/or all components.

Limitations of Backflushing

Backflushing is a powerful tool with a few drawbacks to overcome. First, BOMs must be precise or it is impossible to accurately record production information. However, a high degree of BOM and routing accuracy is required for any agile production planning and control system to be effective. The "garbage in/garbage out" cliche is applicable. There are many good reasons within a world-class environment to have accurate bills of materials — backflushing is just one of them.

Contrary to popular belief, findings reveal that backflushing *increases* inventory accuracy. Assuming the bills of material are reasonably reliable, inventory accuracy will be much better when all issue transactions are done using backflushing. The primary cause of inventory inaccuracy (in a well-controlled environment) is the mistakes people make when entering data. With backflushing there are fewer opportunities for error, so the inventory is more accurate.

A second limitation of backflushing is that it requires a standard BOM to be exploded; there is no room for local variation at the time of manufacture (such as substituting parts). This limitation can be overcome in one of two ways: either (1) the backflushing program has an override capability so that the person entering the completion information can optionally request a component change, or (2) the backflush can be performed from a specific BOM for that schedule. When manual override is allowed, the system will be able to display the bill of material on the data-entry screen, thereby allowing the operator to make any required changes prior to the creation of inventory transactions in the system. When a specific BOM is used, the system will allow for one to be assigned to a particular production schedule. This bill will then be exploded rather than the standard bill for the product.

Extensions of Backflushing

At present most companies use backflushing only to update inventory and production schedules. As the use of backflushing progresses and as agile techniques become more established, backflushing will become more widely used and powerful. The reason for making it more powerful is to eliminate transactions within the systems. Already some companies extend backflushing so that it occurs when the product is shipped to the customer. These are companies that make to order and offer fast delivery. Backflushing reports completions, updates schedules, and records usage of component inventories. If these functions can be done at the same time as reporting shipment to the customer, then one additional transaction is eliminated. Adding backflushing to the program that processes product-shipment information is a straightforward task providing the backflushing program is written in a modular way and can be "called" by the shipment program.

The extension of backflushing the other way, back toward the vendor, has great potential for reducing wasteful transactions. The reasoning is that if the product has been made, then components purchased from the vendors must have been delivered and a receiving transaction can be recorded. The assumption is that the vendor delivers based on inventory pull and that the delivery includes just the quantity needed for production. It is also assumed that production cycle time is short and that no goods-inward inspection and stocking is required. If these assumptions are met, the backflushing program can create a receiving transaction for the purchased components and update the appropriate purchase orders. This approach takes time to implement, and there should be a flag on the purchase order (or item master) showing that the item can have backflushed-receiving.

The next extension of this approach is to eliminate invoices from the vendor. If the product has been made and the components received, then the vendor must have delivered them. If the vendor delivered the components, then the company must pay the vendor. The backflushing program can create an invoice in the accounts payable system. This method is being used by a small number of companies with selected vendors. While there is some resistance to this approach because of the auditing implications, the system can be applied satisfactorily and will save a great deal of non-value-added work in the accounting and administration departments of the company and its vendors.

Inventory Pull

A fundamental difference between traditional and agile manufacturing is the approach toward movement of inventory on the shop floor. Traditional manufacturers invariably "push" inventory through the production process, while agile manufacturers "pull" inventory. The idea of pushing inventory is very much tied into the MRP II philosophy where, owing to long cycle times, complex processes, and material movement from one work center to another, careful and complete planning of the production process is necessary. Shop-floor work orders are created to manufacture fabricated components, manufacture subassemblies, and assemble the final product. Inventory (raw materials, parts, and subassemblies) is allocated to a work order when it is released to the floor, and kits are issued to work centers designated in each stage of the production process. These kits of inventory are "pushed" onto the shop floor based upon the schedule — irrespective of whether the work center can make the product at that time. The thinking has always been that keeping a significant queue of material at a work center ensures that the work center never runs out of work and is therefore highly productive.

Inventory pull takes the opposite view. Nothing is moved to the shop floor until it is needed. In an ideal pull system, a customer order creates demand on a final assembly cell, which then pulls the components required to make the product from feeder cells, which in turn pull their inventory requirements from upstream cells or vendors. There are no queues and no work-in-process (WIP) inventory because everything is pulled according to customer demand.

Elimination of Work Orders

Inventory pull has a profound effect on the way the production plant is controlled. A traditional manufacturer makes use of work orders to plan, schedule, and control the shop floor. The production-planning cycle follows a given pattern. Forecasts and customer orders are reviewed against on-hand inventory and current production plans by a master scheduling program, which creates computer-planned orders recommending the manufacture of additional product. Suggested orders are reviewed by the production planner, who converts them to firm-planned orders. A "firm-planned

---

Eliminating Work Orders

Removing work orders from the shop-floor control system can be done in steps. Select a pilot area or product range within the plant and reorganize the procedures so that, despite using work orders, detailed reporting and tracking are eliminated.

Next, stop using work orders and begin to schedule production using firm planned orders and backflushing. As cycle times and batch quantities decrease, start to schedule using rate-based schedules. Repeat these changes throughout the production plant.

Using firm planned orders as a step toward the elimination of work orders simplifies the transition and makes it easier for people to understand.

---

order" is an order to produce a certain quantity of a product by a specified date, but the order has not yet been released to production. When the orders are needed for release, the production planner converts them to work orders (e.g., shop orders, factory orders) that authorize production on the shop floor.

Orders for fabricated parts and subassemblies are created when the materials requirements planning (MRP) programs are run. MRP reviews the firm-planned orders and work orders from master scheduling and explodes through the bills of material to create computer-planned orders for parts and subassemblies. MRP also creates recommended purchase orders for parts and raw materials that are needed from vendors. The computer-planned orders are reviewed and released in the same way as for master-scheduled items.

A work-order approach runs counter to world-class needs. Work orders assume batch quantities; agile manufacturers constantly strive to reduce batch (or lot) sizes. If batch sizes are significantly reduced, the system will require huge numbers of work orders which become unwieldly to manage. Traditionally, work orders have been used to keep track of the production job through its journey around the shop floor from one work center to another. Cellular manufacturing invalidates this requirement because the entire production process occurs within the one cell. Similarly, with cellular manufacturing and short cycle times, keeping track of where a job is on the shop floor is no longer necessary because there is greater clarity of production and substantially fewer jobs to be tracked.

A work-order system is tied closely to the cost-accounting system. The work order is the vehicle used to collect costs of labor, materials, and overheads as production progresses through the plant. From this data such reports as variance analysis, labor productivity, and overhead allocation are produced. Collection of this data requires a complex system of tracking and control, with the printing of shop-packet documentation that moves with the job and is used for operators to record their times, scrap and yield quantities, and so forth. When production cycle times are short and a small amount of work is on the shop floor at any one time, having such complex tracking and costing systems is not necessary.

Companies moving to agile manufacturing systematically eliminate work orders from production planning and control systems. As cellular manufacturing, cycle-time reduction, short setup times, and lot-size reduction begin to bear fruit, the work-order system is eliminated in favor of an order-free, rate-based scheduling system built around the idea of inventory pull.

Computer systems required to support this approach must be able to handle both work orders and rate-based schedules so that during transition the shop floor is under control. There is a need to have a fully integrated system whereby master scheduling, MRP, and inventory control fully recognize both work orders and rate-based schedules. The same product must be capable of being produced at the same time in the same factory using both work orders and rate-based schedules. This capability is important both during the transition stage and in the longer term because even highly sophisticated JIT manufacturers occasionally use work orders. Work orders may be needed during the introduction of a new product where the detailed tracking of times and costs is required, or when a design-to-order product requires a special project control. The system should have the flexibility to handle both approaches in an integrated way.

Kanban

The *kanban* system was developed by Toyota to control the pull of inventory through the production plant.[1] Thousands of Western manufacturers are adopting the kanban method as they implement just-in-time manufacturing. The idea of kanban, which means "card" or "ticket" in Japanese, is simple. Kanban cards are created for every item made or used in the plant. When an item is needed in production, a kanban is sent to the supplying cell or vendor; this kanban authorizes movement of material to the requesting cell. In its ideal form, the final assembly cell receives kanbans from marketing to manufacture product for current customer sales. The final assembly cell sends kanbans to each upstream cell to obtain the subassemblies and components needed to make the product. The material is moved to the final assembly cell. Each supplying cell manufactures another quantity of the subassembly or component, using kanbans to obtain the raw materials and components required from other cells and vendors. In this way, all material movement on the shop floor is directly required for manufacture, and all activities are synchronized to the needs of the customer. The needs of the final

---

Kanbans

Once rate-based scheduling has been introduced into the plant, it is possible to use kanbans. Initially, use computer-generated kanbans creatd daily (or for each shift) for the current production schedules. Once the process is under control and the kanban quantities are stabilized, it becomes possible to recycle kanbans. A recycled kanban (a card that can be used over and over) removes any daily loading of orders to the shop floor. The kanban system fully drives production.

Later, recycled cards can be replaced by visual methods — like using containers as kanbans, by painting squares on the shop floor, or by having an empty square (or rack) represent a kanban quantity needing replenishment.

Developed over the years by Toyota and other advanced Japanese companies, the kanban system represents a radically different method of production control that is best introduced gradually.

---

assembly cell percolate through the plant and out to the suppliers.

A kanban card has a quantity associated with it. This "kanban quantity" is the production lot size for the item. These lot sizes are small, typically five to 30, and this quantity is held in a standard container for that item. Each cell has a specified number of full kanban containers available in the outbound stockpoint, and these are replenished as the containers are moved to the next cell in the production process. The number of kanbans in circulation represents the amount of inventory of each item available on the shop floor.

There are variations on the kanban theme. A two-kanban system has a kanban for moving material and a separate kanban for authorizing replenishment of the item. The material-movement kanban is returned with the container to the requesting cell and may be recycled next time the product is required. In a single-kanban system, the kanban serves as both the authorization to move and the authorization to make. This procedure is possible when the item moves to the same cell each time and no documentation is required to show where the item is to be delivered. Often the container itself can be used as the kanban; relevant information (such as description, container quantity, and source cell) are printed on the container. The movement of the container authorizes a full container to be moved to the next cell.

The kanban system represents a sophisticated approach to JIT manufacturing. It takes time for the system to be introduced into a production plant — and it cannot be introduced on its own. An important part of agile manufacturing, it requires that quality be excellent and inventory pull be operating — as well as short cycle times, small lot quantities, and cellular manufacturing. Many world-class attributes are embodied within the kanban technique: namely, it is simple to operate, it is easy to understand, it is the essence of inventory pull, and it lends itself to continuous improvement. Inventory levels can be adjusted by merely adding or removing kanbans from the floor; very little paperwork or reporting is necessary for its success. Kanban does not require an entirely repetitive manufacturing process, but the production process must be well understood and fully under control.

The shop-floor control system must be able to support a kanban-style of manufacturing. The system should be able to print kanbans to support the production schedule. These will be production kanbans, material-movement kanbans, and vendor kanbans. In addition, it must be possible to flag an item for kanban control. The system will optionally print more traditional picklists for non-kanban items, enabling a hybrid system to be in use during the transition from traditional to agile manufacturing.

The system should not force the use of kanbans, and the printing of kanbans must be optional and selected by cell, cell range, product, and/or product family. The additional information to support a kanban approach must be available within the system, including kanban quantity, standard number of kanbans, container size, and container type. Inquiries that show the number of kanbans of each item in the outbound stockpoint are helpful and can be used to automatically trigger production on the cell. Algorithms developed to calculate ideal kanban quantities based upon production forecasts may help in setting up a kanban system. In practice, however, manual methods with continuous improvement are probably a better approach.

The Paperless Factory

One objective of agile manufacturers is to simplify production and to eliminate non-value-added activities. The time and effort spent in traditional manufacturing plants on paperwork and reporting is considerable, and the elimination of all unnecessary reporting of information is significant to world-class manufacturing. Such techniques as backflushing materials, using kanban to control the production flow, and eliminating labor reporting enable a company to be largely paperless on the shop floor.

The issue of labor reporting and management-accounting needs is the key to a paperless factory and will be discussed in detail in a later chapter. Replacing paper reporting with electronic reporting does not create a paperless environment because the same activities are merely being done using a different medium. Eliminating the need for detailed reporting is what makes a factory paperless. Agile manufacturing aspects add to the ability to eliminate non-value-added, wasteful data-collection activities. Reduced cycle times and zero inventories are important elements, but underlying them is a process so controlled and understood that detailed tracking becomes unnecessary.

Summary

Execution of production plans on the shop floor differs fundamentally when agile manufacturing techniques are used. The purpose is to minimize inventory, move materials on a just-in-time basis, and eliminate unnecessary transactions. The concept of inventory pull, in contrast to the traditional inventory push methods, is important for minimizing inventory and creating JIT production. Backflushing is widely used by agile manufacturers to limit transactions.

- *Backflushing*
  Backflushing production completions achieves three things in one transaction: it updates finished product inventory, it updates production schedules, and it explodes the components required to make the product and updates component inventories.

- *Inventory Pull*
  Movement of materials through the production plant is achieved by "pulling" components and subassemblies as they are required for manufacturing. A pull system minimizes inventory and ensures synchronized manufacturing.

- *Tracking Lot and Serial Numbers*
  Being able to track lot numbers and serial numbers

from receipt of raw materials through to sale of finished products is increasingly important. Software must support lot/serial traceability throughout the production process.

• *Kanban*
The use of kanban cards as a method of indicating material pull is a valuable technique that must be supported by the computer systems.

CHAPTER FIVE

Order Entry and Customer Service

*T*hree aspects of customer service specifically concern agile manufacturing companies: fast response, flexibility, and a customer-oriented organization. In fact, these three elements work together to provide the world-class service being demanded by customers in the 1990s.

The Customer-oriented Organization

A customer-oriented organization that listens carefully to customers' current needs and long-term aspirations leads to the provision of products and services that truly meet customer needs. A criticism leveled at Western companies in recent years is that they have been so concerned with the needs of design engineers and salaries of senior executives that they have lost sight of the customers. This indictment is too strong, but there is some truth that many organizations have not given customer needs top priority.

Japanese car companies had a very clear picture of what American and European drivers looked for in an automobile in the 1980s — namely, reliability, superior features, and economy (particularly after the "gas crunch"). The "Big Three" car companies in the United States could not provide these features. They made gas-guzzling cars with built-in obsolescence, and every additional feature had to be ordered separately and paid for at a premium. The Japanese companies stole the market. The same story can be told of consumer electronics, cameras, and many other industries. Leaders like L. L. Bean, Disney Productions, Federal Express, and McDonald's learned the importance of listening to customers, understanding their needs, and translating those needs into products and services. This practice is a mark of world-class organization.

Flexibility

Flexibility is a key competitive issue in the 1990s. The 1988 International Manufacturing Futures Report showed that while U.S. and European companies are concentrating their efforts on quality improvement and on-time deliveries, many prominent Japanese and Pacific Rim corporations are changing their operations so as to have a higher degree of flexibility to meet customer needs.[1]

Three prime aspects of flexibility are mix flexibility, volume flexibility, and design flexibility. "Mix flexibility" is achieved by offering short lead times so that product mix can be changed very quickly in accordance with customer demand. "Volume flexibility" is the ability to economically manufacture the quantities required by the customers when they need them. This system entails being able — at the drop of a hat — to change production volumes and to make as much or as little as is required.

"Design flexibility" is concerned with the ability to introduce new products or major design upgrades quickly and effectively. The speed with which a company can introduce new and upgraded products can be the difference between winning and losing in the marketplace. The introduction rate of new products has been astounding in recent years, and this trend will continue. Agile companies have learned to shrink the time-to-market through good design practice, value engineering, concurrent engineering, and a streamlined process of introducing new products into production.

Responsiveness to Customer Needs

Fast response is concerned with lead times and production cycle times; companies can obtain a competitive edge by providing products and services more quickly. A traditional company will achieve fast response by holding inventories of finished products and servicing customers from inventory. Agile manufacturers develop fast, flexible production processes that can serve customers needs for quick response while still allowing the product to be made to order. Such a company can give customers the fast response they need while at the same time maintaining very low inventories.

Another aspect of fast response is being able to provide custom-designed products. A traditional manufacturer either has a limited number of products available from inventory or offers long lead times to provide customized items. An average company declares it impossible to quickly provide custom products. World-class manufacturers study and improve their processes until delays and problems are eliminated.

A much wider range of product has become available in recent years because customers are requiring that more specialized needs be fulfilled. An everyday example is the Coca-Cola Company, which for several decades packaged Coca-Cola in bottles and cans of various sizes. During the

---

Using TQC Techniques

Total Quality Control (TQC) techniques — including cycle-time analysis, Pareto charts, statistical process control (SPC), fishbone diagrams, and quality circles — are just as valuable for improving clerical processes such as order entry.

Study the speed and accuracy of order-entry and customer-service processes. Use fishbone diagrams to analyze the causes of error and delay. Use SPC to determine the degree of variation and if the process is under control. Use quality circles to establish a total-quality, continuous-improvement approach.

---

past ten years it has produced diet Coke, caffeine-free Coke, caffeine-free diet Coke, cherry-flavored Coke, as well as "classic" Coke — all packaged in a wide range of bottles, cans, and dispensers in multiple languages. An agile manufacturer is aware of the customer's desire for variety and specialization and fills that need with products and services that cannot be provided by traditional methods.

Lehigh University's Iacocca Institute issued an influential study on a 21st-century manufacturing strategy. It presents a vision of the "agile manufacturer" whose entire business strategy and competitive edge centers on flexibility and fast response. The report postulates an automobile company that can deliver in three days a vehicle that has been custom-designed in the sales showroom *by the customer* (who is called a "prosumer" — not a consumer) using computerized design techniques and highly integrated communications between showroom, plant, and customer. Of course, this vision is many years away. While it is not currently technologically feasible, it is the direction toward which leading companies are headed.

A rapid level of response to customer needs while maintaining just-in-time (JIT) inventories of finished product requires considerable sophistication and skill with agile techniques. This response level cannot be achieved overnight. It requires a concerted, dedicated, long-term implementation of all aspects of world-class manufacturing where one improvement builds continuously upon another until zero inventory, high quality, and JIT production become feasible. Such dedication to customer service and long-term improvement sets apart a world-class corporation.

Software for Customer Service

Analysis and elimination of non-value-added activities is as applicable to distribution and service functions as it is to manufacturing. Determining the causes of poor service and the elements of lead time is very important. When analyzing the lead time offered to customers, many companies find that most of it is nonproductive, consisting of the order waiting to be processed, checked, or actioned. Orders are held up in order entry, credit checking, inventory control, production planning, engineering, the stockroom, packaging, or dispatch. The same story applies to invoices that often take many days to produce and send out.

Studies show that most companies can cut lead times by 50 to 75 percent by just analyzing the flow of information and removing roadblocks in paperwork flow from order entry to dispatch. An example is taken from a company making wire mesh for window screens. This mesh comes in a variety of standard sizes, gauges, and colors. It formerly took four days from the customer placing the order to the time the company could quote a delivery date. This four-day period consisted of handwriting the order and sending a copy to the stockroom, a copy to production planning, and a copy to accounting and credit control. Writing orders by hand often took more than a day because orders were done in batches and the order-entry people were overloaded. Additional time was needed to circulate copies to each department using the company's internal mail service. Credit checks, stock control, and production planning were done simultaneously and independently. Often the scheduler would make a last-minute production revision to squeeze in a new order only to find out later that the customer had been put on credit hold and that the new order had to be removed. All this paper-work constituted waste to be eliminated.

Most of these problems are caused more by poor management of the customer-service function than by inadequate software. However, a well-integrated order-entry and production-planning computer system can remove many obstacles and can streamline the entry, checking, and processing of customer orders. Although no computer system can make a poorly organized company or department efficient, great improvement can be made by simply understanding the process and eliminating the causes of delays and errors. Total Quality Control (TQC) can be well applied to a customer-service organization. Analysis tools used by agile manufacturers (statistical process control, cellular organization, quality circles, Pareto and fishbone charts) can be readily adapted to the office environment. The concepts of JIT, small lot sizes, and short cycle times can be applied to the flow of orders, product dispatch, and invoicing through the customer-service process.

The Integrated System

A well-integrated computer system for customer service can help reduce processing times and can increase accuracy — accuracy both of the order-entry process and of answering customer inquiries. The access to valid and timely information makes an integrated system valuable. This information includes customer details, credit checking, inventory availability, availability of alternates, manufacturing available-to-promise, and automatic pricing and discounting.

When an order is entered, the system checks the customer's record and obtains all information about the customer's delivery address, contact name, special instructions, and terms. The system can check credit history and rating to confirm whether the order should be processed immediately. If the customer fails credit checking, the order can still be taken, but product dispatch is prevented until the credit issue is resolved by the person responsible for accounts receivable. The elements and sophistication of the credit-checking process vary according to the needs of the company and the market. A straightforward check of the total accounts-receivable balance for the customer, including the new order, against a predetermined credit limit is often sufficient. More complex credit checking takes into account the number of days the customer is late in paying current invoices, the customer's payment history including average lateness, and the customer's status and importance to the company.

Inventory checking requires that the system interrogate warehouse records to determine if there is enough inventory to meet the customer's needs. This action, while primarily of value to companies that make and dispatch to stock, is also useful to the make-to-order manufacturer because some finished-goods inventory invariably builds up. A company usually has more than one warehouse or stockroom, and the computer system needs to know which stockrooms are available for shipping inventory. The order-entry process will review inventory available within these stockrooms. When inventory is not immediately available, the system can automatically show other locations that have available inventory of the product. In this way, order-entry people can source the product from another location, such as an adjoining region.

Available-to-Promise

When inventory is not immediately available to meet a customer order, providing an accurate promise date for delivery is essential. This date can be obtained through the "available-to-promise" calculations. Available-to-promise takes inventory availability one step forward; instead of just taking account of currently available inventory, the system also has knowledge of production schedules and purchase orders. This information can be used to calculate how much inventory will be available each day in the future. Availability is calculated by adding the expected supply quantity (from purchase orders or production) to the on-hand quantity and subtracting any demand (such as customer orders, transfers, production component requirements, and shelf-life expiration quantities). The result is the amount of inventory that will be available each day, based on current plans. The system reviews these figures and determines when the customer's order can be fulfilled. The promise date is set according to the future availability of product.

Some systems take the ideas of available-to-promise a step further by allowing the automatic allocation of an individual customer order line to specific production orders or purchase orders. This form of allocation is called a "set-aside." As product is manufactured or received from the supplier, inventory is automatically earmarked and set aside for that certain customer order. Even more helpful is a system that not only sets aside order lines to specific production, but also sets aside quantities of inventory to a customer order ahead of time without a specific order being raised. Orders, when received from the customer, are then fulfilled from a quantity of product that has already been set aside. This system is particularly useful when blanket or contract orders are received from a customer and call-offs are made daily for the actual delivery quantities. The total blanket-order quantity can be set aside against future planned production even though the daily order quantities are not known.

Pricing

Pricing is the most volatile aspect of software design for both agile and traditional manufacturers because each company has its own way of handling price and discount issues. Considerable flexibility is required in the way prices are calculated.

In "customer-specific pricing," the price is agreed upon ahead of time with the customer. Every time the customer places an order, the same price is used. A useful feature is to have date ranges associated with customer-specific prices so that the price reverts to a standard when the contract expires.

Order Entry

The most most important feature of an order-entry system for world class manufacturing and distribution is that it should allow orders to be processed quickly and correctly. A complex and confusing order system will result in errors, delay, and customer frustration. Order-entry systems are crucial to effective customer service. Do not overcomplicate the process — design it for simplicity, clarity, and speed of service.

Many companies have complex and convoluted pricing and discounting policies that often are necessary, driven by customer need, and the result of many years of tradition. Simplicity may require a revised approach to pricing and discounting.

Price lists can be set up for a specific customer or range of customers. This method is similar to a manual system because the company sets up several price lists, and each customer is assigned to one specific price list such as retail or original-equipment-manufacturers (OEM) prices. Price lists can contain the prices themselves or an uplift percentage from a base price; the uplift can be either positive or negative. The price list (and/or the base prices) can have effectivity dates, thus allowing multiple price lists using the same price-list number but having different date ranges. This list can be used for annual price increases, special offer periods, and commodity pricing where price varies frequently.

"Quantity pricing" refers to a set of prices based on the quantity the customer is ordering. The system automatically selects the price from the price list according to the quantity being purchased. Typically, of course, a higher quantity attracts a lower price.

"Matrix pricing" allows the selection of the individual price from a series on the price list according to a matrix table that takes account of customer type and product type. The price list may contain ten or fifteen prices (prices 1, 2, 3 and so on) for each product. The decision of which price to choose from the list is made by looking up a price matrix table containing all product types and customer types. For example, the customer type may indicate geographical location and the product type may show whether the product is perishable or nonperishable and requires special transportation.

"Manual pricing," which is always needed, simply allows the order-entry person (or the person with the relevant authority) to manually set a price according to company procedures or negotiation with the customer.

There are many variations on pricing methods, and some companies have very unique pricing methods that are an important part of their marketing strategy. Individual pricing plans must be built into the order-entry system. For an integrated system to work effectively, pricing calculations must be done on-line at the time the order is entered. Thus, the order-entry person can review prices as he or she goes along and, if necessary, allow prices to be given to the customer immediately. Some standard systems provide an option to enter orders in large batches and for the system to calculate the prices, discounts, and credit-checking information at a later stage. Such features are included when the machine is slow and order volume is high, allowing large numbers of orders to be entered quicker and the calculations done off-line. Occasionally this feature is useful, but real-time on-line is the best way to operate an order-entry procedure.

Discounting

Once a price has been established, customer discounts can be calculated. Discounts can be established for the individual order line based on quantity. They can be calculated from the total quantity (or other factor) of all lines or from the total quantity purchased by the customer to date, or by agreement with the customers.

"Line-item discounting" is simply a matter of discount (or uplift) percentages being selected from a set of discounts according to quantity purchased. The discount table contains not only discount percentages, but also the quantity breaks that divide the discount rates. Frequently, the discount will be expressed not as a single discount percentage but as the combination of two or three discount amounts. These discount percentages represent different aspects of the discount (such as for location or for marketing reasons) that must be shown separately. Often these separate discounts are posted to different accounts on the ledgers to allow for the separate analysis of different types of discounts.

"Total-order discounts" also work from the quantity purchased. However, instead of considering a single line quantity, the entire order quantity is used to determine discount prices. Being able to do total discounting by product type, instead of just by entire order quantity, is important. This method allows for different total-quantity discounts to be applied to different types of product. Total-order discounts do not have to be established from quantity purchased; the rate can be based on weight, volume, number of cartons, number of pallets, and so on. Any factor that serves as a method of identifying an advantageous combination of products can be used to determine the amount of discounts granted.

The "purchased-to-date" method of assessing discounts is not concerned with the size of individual orders. Rather it adds together all product purchased by a customer during the last twelve months (or more) and assigns discount percentages that improve as total purchases increase over time.

Product Searches

Traditional computer systems rely on the order-entry person knowing the item number of the product being purchased. There may be catalog numbers the customer gives to identify the item or the order-entry person may have good product knowledge, but one way or another the item number has to be determined prior to entering the order. Agile manufacturers and distributors recognize that their systems require greater flexibility. Customers do not necessarily know the catalogued item numbers; customers may wish to place orders using their own company's assigned identification numbers; there may be industry-specific, generic item numbers for some products; products may be identified by characteristics, not numbers; or there may be so many products that order-entry people cannot possibly know the numbers.

There are a number of ways around this problem, most of which still rely on products having unique identifying item numbers that are used by the system to keep track of that product. However, it is not necessary for these numbers to be widely known and used. The simplest solution to this problem is having cross-reference tables for the products, an especially good system when customer-specific or generic part numbers are in use. A table is set up which cross-references the item numbers, and the order-entry system has the functionality to search for the item through the cross-reference tables. The cross-reference tables may be customer specific so that the customer's part number is easily accessed. These customer cross-reference tables can also hold additional information like the certification code, specification numbers, or material-safety data-sheet information required by that customer.

The advent of relational-style data bases makes possible the provision of a very large number of search criteria for every product. This model allows the customer to describe the product instead of quoting an item number. The order-entry person enters the key words used to describe the product, and the system lists all products on file that match this description. Typically, this kind of product search has a table of around 20 words to reference the product. These words may be assigned automatically from the product's description and/or manually added by the people who set up the product on the system. Products listed by the system are those whose word-search table contains all words entered by the operator. Order-entry people can then narrow down the search until the required item is determined.

A useful extension of this approach is product-specification tables. These tables present characteristics of the product in a standard form. For example, a paint manufacturer may have a specification table containing color, base, finish type (such as matt or gloss), and package size. The order-entry person enters the characteristics of the product required, and the system searches the tables and lists all products that match. This procedure is similar to the search table described earlier, except the key words that identify the product are arranged in a structured way on a user-defined table. This method is even more powerful when specification tables can be changed for different product groups. A specification-table number is assigned to each product, and the order-entry person tells the system which type of product to look for. The screen lists the specification headings for that product group, product characteristics are entered, and matching products listed.

This system provides a simple, structured method of locating products being ordered without expecting customers to know company item numbers. It is particularly useful when there are many variations of the final product. For example, a clothing manufacturer will have multiple sizes and colors for the same item, the order-entry person can define the product, and all the different sizes and colors can be quickly identified. Such an approach is also useful for materials whose characteristics are not known until the product is manufactured. Specific characteristics of each product batch (e.g., viscosity, dimensions, concentration) is analyzed and the results entered into the product-characteristics table. When a customer places an order and specifies the characteristics required, order-entry people can search for products with characteristics that match those required by the customer. Specific batches or lot numbers can then be allocated to that order.

Customer-specific Product Lists

Customer-specific product listing, another valuable approach to the identification of products in some industries, is useful when customers purchase the same items on a regular basis. Items may be purchased by that customer only (name-brand products, for example) or they may be standard items ordered over and over. A good example is a supplier to the food retail business where customers (markets and stores) buy products on a weekly or daily basis and purchase different quantities of the same items every time.

Instead of the order-entry person having to enter item numbers every time, the system can bring up a list of products the customer typically buys — a kind of electronic order form — and require only the entry of the quantities. In addition, the electronic order form can show such information as recent sales history, average demand, and prices. This quick, accurate, and simple plan can also be used by order entry to prompt the customer to order items they may have overlooked.

Lists used for this method of order entry can be created in different ways. One model has the system automatically keep track of what the customer has ordered in recent weeks and display those items. The user specifies how far back the system should look when recording a customer's buying habits, and the system will automatically create and update the table. Another approach manually creates customer-specific product lists. This mode is more time-consuming and complex, but does enable the user to apply some additional marketing knowledge to the process. For example, short-term promotional items can be removed from the list and alternate products (which the customer has not previously purchased) can be added. A combination of both automatic and manual is ideal, but most standard order-entry systems provide one or the other.

Types of Order Entry

Order-entry systems required by agile manufacturers vary enormously because their products and customers vary.

Approaches to Order Entry

The challenging paradox faced by companies moving into agile manufacturing is to provide customers with greater flexibility and a faster, more complete service while at the same time simplifying procedures associated with providing this service. This paradox manifests itself clearly in the order-entry process. The increasingly complex demands from customers must be met with increasingly simple systems.

Start by recognizing that a standard order-entry system must be tailored to your company's needs and achieve the following:

- provide functionality that focuses on customer needs and world-class objectives
- fearlessly remove unnecessary features
- provide minimum transactions in the order-entry and fulfillment processes
- integrate the processes to minimize lead times
- move away from single orders to blanket orders with daily call-offs
- be clear and understandable to everyone
- continuously simplfy the system as WCM methods simplify the process

---

Most tend to be complex in the way they work, providing many features and override capabilities. While such complexity is sometimes necessary, often it has developed over time and no longer reflects the true requirements of the business.

A company moving toward a world-class approach seeks to eliminate waste, shorten lead times, and provide better customer service. Simplification is a key element to the achievement of these objectives, and examining the process of taking and filling orders to eliminate wasteful activities is important. It is important to establish the sources of confusion, mistakes, and delays so that these problems can be resolved and eliminated. These clerical activities should be treated no differently than those on the shop floor when it comes to waste elimination and process improvement. They are like all other procedures. Establishing value-added and non-value-added activities is important, as is reducing queues, establishing cells of activities, breaking down departmental barriers, providing cross-training, establishing quality circles, and fostering continuous improvement.

A good example would be the integration of customer service and the accounts-receivable department. Order-entry and credit-control procedures are intimately related. Creating work cells within these functions with cross-trained personnel can eliminate considerable waste and delay in customer service. For instance, a manufacturer of fashion shoes, prior to taking an agile approach to its order-entry and fulfillment processes, had a delivery lead time of ten to fifteen days. The procedures used to service "rush" orders (which had a five-day turnaround time) were disruptive because the "rush-order" people would often take items that had been allocated to other orders. This action resulted in considerable confusion and expediting orders at the last minute. An examination of the order-entry and fulfillment processes resulted in the establishment of customer-service cells where everybody could do all the tasks required to fill the customer orders. Together with a new order-entry system, credit checking and inventory allocation was performed on-line and resulted in the customer lead time being reduced to two days. This is a real success story. Sales have increased as a direct result of improved service because the company is now more responsive to the changing needs of the fashion market and can meet retail demand during the critical seasons.

As with all aspects of agile manufacturing, computer systems should have the flexibility to accomodate changes taking place without being complex and confusing. As simplification and standardization of processes is introduced and the company becomes more responsive to the customers, the systems must reflect this new simplicity. Even when the simplification of processes, products, pricing, and customer service has progressed, there remains the need for more complex order-entry procedures to handle unusual orders or newly introduced products. This flexibility must be in the software and not obscure the simplicity of the regular process.

High-Volume Standard Products

A company selling high volumes of standard products needs a simple and highly automatic order-entry system. Order-entry people typically will not require as much product knowledge as that demanded by a company with a more complex range of products. Activities like automatic pricing and discounting, inventory allocation, alternate products in case of stock outs, and printing of confirmation notices, picklists, shipping documents, and invoices should be simple and straightforward.

Product-search capability and customer-specific product lists will be helpful, as will automatic telephone-marketing features. For example, some telephone systems can provide the telephone number of the party calling for customer service. This feature can be integrated with the order-entry system so that customer information is already on the screen when the call is answered. Time and effort is saved — and more importantly — a highly professional image greets the customer.

Specialist Order Entry

When people taking orders and interracting with customers are knowledgeable of the products and have a wider range of authority, a more flexible order-entry system is required so that specific needs and unusual features or requirements can be added. Flexibility is needed in each area of the order-entry process. The order-entry person needs the ability to easily change all aspects of the order, including the ship-to address, terms of the sale, shipping information, prices, price lists, discounts, accounting information, delivery dates, warehouse picking sequences, commissions, and other information.

Of importance to an agile manufacturer is the time needed to service a customer. Information must be retained not only on when the order is committed for delivery but also on the date the customer really required the item. Most companies keep track of delivery date against promise date; few companies keep track of how that information matches the customer's stated needs. Another piece of important information records the date and time the order was taken from the customer. A simple and effective performance measure for a fast-turnaround supplier evaluates the cycle time from taking the order to shipping the order (or better yet — to customer receipt of the order). This information is easy to gather and easy to report yet not used by most organizations.

Call-Off, Blanket Orders, and Automotive Releases

Agile manufacturers are moving increasingly toward simpler methods of ordering products from suppliers. A supplier that is also a world-class organization will change order-entry procedures to simplify the receipt and processing of orders. A blanket order, the first step toward this kind of simplification, is created for the total expected requirements for an item over a time period, usually one year. This blanket order establishes the terms and conditions of the transaction but does not specify delivery dates and quantities.

When the items are needed, the customer "calls off" the requirement from the supplier giving the delivery dates and quantities. This procedure eliminates the need to raise a new purchase order every time an item is required because the purchase order has been negotiated and agreed to in advance. In addition, calling off can be done by the shop-floor people who need the parts rather than having to involve the purchasing department. The arrangement reduces administration and paperwork and gives operators the authority to bring in materials and parts on a JIT basis.

This call-off process can take place in many different ways. It can be a formal receipt of a call-off document, a faxed copy of a call-off document, a telephone call, receipt of a *kanban* card, or an electronic or automatic signal. The objective is to eliminate as much of the order-entry process as possible. In the early stages of introducing simplification procedures with a customer, some accounting checks and balances must be retained. However, as the relationship develops between the two companies, many administrative procedures can be eliminated altogether.

The ultimate objective achieves the elimination of all administrative procedures. Products can be manually called off using kanbans (or something similar) without entering the computer systems at that point. Order-entry and shipment information can be updated when customers confirms that they have used the product through a backflushing process. If the customer pays at the time the product is used and this payment is notified through the backflushing process back to the vendor, then the entire process can be achieved without any data-entry steps until payment is received. Of course, this ideal is difficult to achieve but should be the objective because it minimizes the non-value-added administration process. Using electronic data interchange (EDI) can also simplify this process, but it must be remembered that EDI automates procedures that potentially could be eliminated altogether. Do not be tricked into thinking that your company has minimized transactions by using EDI when it is merely automating waste. An indepth look at EDI is presented in Chapter Six.

The concept of release against blanket orders has been in use within the automotive industry for some time. First-tier automotive component or subassembly manufacturers often place long-term orders with their suppliers and then call off the materials as they are needed. The call-off may be daily, twice weekly, or weekly depending on the type of product the vendor supplies and the proximity of the vendor to the customer. When the customer is close by, greater flexibility is possible. Many automotive companies use the "fab and raw" concept of risk sharing with their vendors. When a call-off is conveyed to the supplier, the customer also provides a forecast of future requirements in order to assist the supplier in materials and production planning. However, the call-off quantity is the only firm requirement.

The agreement between the supplier and the customer is set up so that the customer will pay for the fabrication of components based on forecast up to an agreed number of days in the future. Similarly, the customer will pay the supplier for any raw materials purchased to support the forecast up to an agreed number of days in the future. The "raw" time is longer than the "fab" time and allows suppliers to confidently plan production and purchase materials even if the customer fails to call off the items. The fab and raw times typically are five and ten days although the forecast of expected requirement can go far into the future. This procedure is typical of close, cooperative relationships that exist between customers and vendors in an agile work environment.

Kits

Many companies find a kit feature to be useful in their order-entry systems. If the company manufactures standard products and then groups these products together into sets when the product is sold, preplanned sets can be available within the order-entry system. For example, a company manufacturing (or distributing) scientific equipment will manufacture not only the products themselves but also a range of accessories, attachments, cases, and consumable items. When customers place orders, they will usually buy these accessories and consumable items along with the principal product.

The order-entry system will allow for the entry of kits and, when an order is placed by a customer, the order-entry person enters the kit identifier and the system generates order lines for the entire kit. The kit features must allow pricing for the entire kit, superseding the prices of individual items. If kits are already made up and held in inventory, then these kitting features provide just a listing of the items contained within the kit. If the kits are made to order, then the kitting features will trigger the production of a kit and also do the inventory transactions required for component items. The kits need to be flagged to show how the sales and bookings analysis will keep track of the order. Some companies will want to keep a sales history of kits, whereas other companies will want to track the sales of individual items irrespective of whether they are sold separately or as part of a kit.

The use of kits can simplify order entry and reduce the number of transactions required to enter an order, thus saving time and effort and reducing waste and data-entry errors.

Acknowledgment

Once the order has been entered, some customers require acknowledgment. Acknowledgment is a wasteful and unnecessary task and should be eliminated as soon as possible. Its primary rationale is to assure customers that their orders have been placed correctly. This traditional thinking is linked to concepts of accounting control and low quality. Agile manufacturers seek to establish close working relationships with customers and vendors in order to eliminate any need for these kinds of procedures.

The close relationship between some agile companies and the customer allows the company to determine the order quantities. Instead of the customer placing a purchase order with the supplier, the supplier has visibility of the customer's inventory levels and planned usage; the supplier calculates the quantities the customer needs, places the order, and ships it. Under such circumstances the customer will often require an order acknowledgment that communicates the order quantities. Increasingly, this acknowledgment process is achieved using an EDI transaction.

For this approach to work successfully, the supplier must have detailed information about the sales and stock levels in the customer locations. If this information can be readily included and brought down into the supplier's planning and control systems, the company's material requirements planning (MRP) or distribution resource planning (DRP) systems can review the information and create replenishment orders on behalf of the customer.

This procedure of having the supplier determine the order quantities is not new. Bread companies have been doing this for many years when restocking supermarkets. Rather than delivering from an order, the delivery people take bread into the market and refill the shelves according to how much was sold on the previous day. They are also careful to remove any products with expired shelf lives.

Picking, Packing, and Dispatching

Once an order has been entered, the task of filling that order and shipping it to the customer begins. An agile manufacturer will seek to minimize all administrative activities and material movement. Many traditional companies and software systems had elaborate methods of acknowledgment, allocation, picking, confirm pick, staging, packing, and dispatch. World-class manufacturing and distributing seek to simplify and eliminate these processes. A company that manufacturers to order on a JIT basis will not require the picking, staging, packing, and dispatch processes because dispatch will be a part of the production process with no additional paperwork required. In most instances, product dispatch will need to be recorded because at that point the product becomes the property of the customer.

Picking, Packing, and Dispatching

The move to agile manufacturing necessitates dismantling the complex picking, packing, and dispatching processes used by traditional companies. A complex dispatching process is unnecessary when finished goods are low and lead times are short.

Start by eliminating multiple steps. Instead of picking, staging, confirmation, and dispatch, move to a single-step pick and dispatch. As finished-goods inventory is eliminated, begin to dispatch directly from completed production without needing the inventory and picking step. Later, eliminate the production-completion step and record completions and backflushing through the dispatch transaction.

It is important to minimize transactions. In some cases the dispatch program can also record production completion and backflushing. However, rather than recording a production completion and then entering a dispatch transaction, these two tasks can be combined into one program. This allows a single entry to record the completed quantity, back-flush the components, update the production schedule, record the dispatch against the customer's order, and create an invoice entry into the accounts-receivable system.

Achieving this degree of simplicity is unusual. Most companies do have a separate dispatching process. After manufacture, the production completion records the items as a finished product (i.e., backflushes the components) and the item is made available for dispatch. When the product is dispatched, a separate transaction is entered recording the dispatch of the item from finished goods, updating customer-order information, and providing invoicing information.

Finished-Goods Inventory

A company without significant finished-goods inventory does not require any kind of warehousing process because assigning an item to finished goods is enough to identify and locate it. However, many companies do maintain a finished-goods inventory stockroom and therefore need to track what is in it.

Finished-goods inventory is necessary when the company's products have significant seasonality and have to be made ahead of time to offset capacity constraints at peak periods. Similarly, a company may need to hold some finished-goods inventory to accommodate erratic customer demand. Eliminating all such inventory through the use of JIT manufacturing and production flexibility is ideal but may not be feasible in the short term.

The production-completion program not only can record the transfer of the item into finished goods, but can also assign a stocking location where the item is stored. Frequently, the primary bin location for an item can be used, but companies using a more random storage method will want the program to locate an item according to predetermined rules. As a company moves toward world-class status, such complex warehousing operations must be eliminated in favor of simplified systems. While random storage warehouses with automatic assignment of bin locations are often necessary for a high-volume distribution organization, they should be eliminated as soon as possible.

Soft Allocation and Hard Allocation

Inventory allocation is used to earmark specific inventory for a customer order. These products can be used only to fill the customer order for which they have been allocated. There are two kinds of allocation — soft allocation and hard allocation. "Hard allocation" earmarks one (or more) individual occurrences of an item. For example, if the company has fifteen of an item in stock and these fifteen are located in different places within the warehouse or factory, hard allocation can allocate a particular piece against a customer order. The individual piece in that stocking location is the one that will fulfill the customer order.

"Soft allocation" does not identify particular pieces for allocation but simply keeps track of how many are available. If the quantity of fifteen is available and a customer order requires one, then soft allocation will show that only fourteen remain for future orders. When the customer order is fulfilled, any one of the fifteen may be used. Soft allocation concerns itself only with the available quantity and not with the particular piece being used.

Hard allocation is useful when the item being ordered has specific characteristics required by the customer and the order-entry process selects a particular piece from the available stock. Hard allocation can ensure that material is issued on a first-in/first-out (FIFO) basis, or by date or lot-number sequence. Other allocation rules used include allocating the entire quantity from a single lot number (if possible), allocating items that are physically close together to prevent unnecessary travelling, or allocating items in small quantities so as to free up bin locations for future use.

An order-entry system must provide soft allocation. Only through soft allocation can the available inventory level of an item be shown. Hard allocation, not usually necessary in an agile environment, is used by traditional manufacturers to earmark items that cannot then be used by other orders or processes. This requirement assumes that the item will be hard allocated for a considerable time. Agile manufacturing's concern is to fill orders immediately — not earmark items for future use.

Hard allocation sometimes identifies items that cannot be used for regular shipment against sales orders because they are beyond shelf life, or have failed quality control, or are required for use by engineering or marketing. In such situations the material preferably is assigned to a segregated area, either physically or just within the inventory-control system. This simpler approach to resolving the issue does not require any complex hard-allocation features.

Usually, allocation is used for currently available inventory. However, it can be used to allocate future available inventory. The term "reserved inventory" is commonly used instead of "allocation" to show that the item is not physically available. Inventory can be reserved from future production orders, schedules, or master schedules — or, in the case of purchased items, reserved from a specific purchase order. These inventory-reservation methods are again more prevalent in traditional manufacturing where long lead times and detailed production control require the creation of future orders against which reservations can be made. An agile manufacturer will steer away from complex inventory-reservation facilities, preferring to manufacture on demand and with short lead times. Short lead times and production flexibility remove the need for both hard allocation and reserving future inventory.

Picklists and Confirmation

A company using finished-goods inventory to meet customer needs often utilizes a picklist to identify items to be removed from the stockroom and dispatched to customers. A picklist authorizes the stockroom personnel to move products and instructs them where to find them. This plan is particularly important when hard allocation or allocation rules are used because stockroom personnel need to know which specific item to use to fill an order.

There can be a separate picklist for each customer order to be shipped — or a "consolidated picklist" containing a series of orders that is sorted into picking-location sequence can be used when the stockroom personnel pick many orders together and need an optimum travel route through the warehouse. Consolidated picking requires a staging process where individual orders can be assembled from picked inventory.

Picklists, or their electronic replacement, are required in a complex distribution environment but should be eliminated by agile manufacturers. The ultimate objective is to eliminate finished-goods inventories and make to order, in which case a picking process is not required. If finished-goods inventory is required, a simpler system can eliminate

---

Inventory Accuracy

Chronic inventory-accuracy problems stem from poor inventory-control procedures. Traditional companies use an annual physical inventory to correct errors. However, while pleasing auditors, a physical inventory is subject to many transaction mistakes and often leaves inventory accuracy worse than before.

Many companies are moving to cycle counting. Cycle counting uses an ongoing physical stock count in which a few products are counted each day and errors are corrected. Parts are selected logically — based on value, amount of movement, and so on.

While providing a more accurate inventory level, cycle counting is still a wasteful activity to be eliminated eventually. It is best used as a continuous-improvement method. Detected errors are thoroughly investigated and their causes permanently resolved. This approach perfects the process and eliminates the need for cycle counting.

---

the need for printing a separate picklist. Products can often be picked directly using the sales order, thus eliminating the need to print the picklist and providing a more direct, fast turnaround of the orders. Also, a careful design of the stockroom so that the location and availability of products is visible and easily understood can eliminate the need for picklist processing. Training stockroom personnel in problem-solving techniques can significantly eliminate waste in the process.

Complex warehousing operations include a confirmation-of-pick transaction. When the picklist is printed, no inventory transactions occur; the confirmation entry creates the transaction that updates inventory after the material is picked. The confirmation validates that the parts and quantities on the picklist have, in fact, been picked before updating inventory balances. A picking confirmation step is very wasteful. It is there because the company lacks confidence in its inventory accuracy and will not update inventory until the items have been physically located.

An agile manufacturer eliminates confirmation-of-pick transactions and works toward accurate inventories. Inaccurate inventories are a quality problem, and a *production* quality problem is not solved by performing more inspection. Quality cannot be "inspected into" a product — production quality is created by perfecting the production process through the use of world-class production techniques and employee involvement. Similarly, accurate inventory records are created by simplifying the process, minimizing transactions, using fail-safe reporting methods, and involving the people doing the job in the solution to the problems.

Confirming Shipment and Receipt

Most companies require entry of a transaction that reports the shipment of product to the customer, but few companies confirm receipt to the customer. Customer receipt confirmation is only prevalent in industries dealing with toxic or highly regulated substances which require careful control and tracking.

The shipment transaction in most companies — agile or otherwise — updates finished-goods inventory, updates the sales order to record the shipment of the product, and records the point at which the invoice transactions are created. Often the invoices themselves are printed at a later stage but the transactions containing the invoice information are created through the shipment-entry program. A JIT manufacturing company may also use the shipment-confirmation program to report product completion and to backflush the components and raw materials used to manufacture the product.

Shipment confirmation should work on an "exception" basis. The person reporting the shipment does not have to enter all the detailed information line-by-line; he/she can report the shipment of the order number, and the system creates the shipment transaction for each individual line within the order. The person reporting the completion can override the quantities or the product shipped on an exception basis, changing only those items that differ from the customer order. Some systems allow entry of a picklist number, and shipment transactions are created for all the picklist items.

The requirement for a shipping confirmation can be eliminated when a company has a very close working relationship with one or more of its customers. If the company provides components or raw materials to a customer with a JIT manufacturing process, then component usage can be reported as part of the backflushing of the components when production completions are reported by the customer. If the products have been made, then the components must have been used; if the components have been used, then they must have been called off from the supplying company. If they have been delivered by the supplying company, then an accounts-payable invoice can be created to arrange for payment of the item. While this level of trust and cooperation does not commonly exist between customers and suppliers, it is the objective of an agile vendor relationship. The savings in administration costs and transactions are enormous if this level of integration can be created between customer and vendor.

Receipt confirmation is, in fact, valuable information for a world-class company concerned about customer service. Recording and measuring the company's success at product shipment is a useful performance measure, but product receipt is a more pertinent measure from the customer's perspective. The problem is that this information is difficult to gather. Establishing an elaborate method of recording receipt at customer sites runs counter to the goal of eliminating transactions and complexity.

Receipt information can be collected using reply cards — but this method is time-consuming and getting customers to return them is difficult. A company using a trucking company to deliver products can ask the trucking company to report back the date and time of customer delivery. Some trucking companies provide this amenity as a standard service. However, even this information is time-consuming to manually enter into a computer system. Some trucking companies provide this feedback information on computer files that can be automatically read and analyzed by the company's system. Alternatively, if the customer can use electronic data interchange (EDI), receipt information can be transmitted back automatically to the supplying company through EDI transactions. Instead of being used for reporting shipments, current receipt information is valuable data for measuring performance.

Material-Safety Data Sheets

There is an increasing need in the United States for the inclusion of material-safety data sheets (MSDSs) when products are shipped to customers. This is due to the additional regulation of toxic and environmentally harmful materials by the federal, state, and local governments. Government agencies of many other countries also are taking a more active role in the regulation of hazardous materials. The European Economic Community (EEC) has an agency in Brussels that provides controlling legislation for hazardous materials.

Material-safety data sheets provide information about the substances contained within a product; they alert the user to the dangers, provide safety information, give instructions for disposal of the material, and define actions to be taken in times of emergency such as a fire or an accident. In addition to providing this information to the customers, companies are also required to provide their own employees with information relating to hazardous materials used within production plants. Most companies do not need computer systems to produce the MSDS because, although they manufacture many products, there are few hazardous materials contained within these products and MSDS procedures can be handled manually. However, as the regulations become more exacting and as companies use more hazardous materials, the need for computer-based systems to disseminate this information in accordance with government legislation and the company's own policies becomes more important.

An MSDS system contains a data base of information about the hazardous products used by the company. Owing to the repetitive nature of the information printed on MSDS reports and labels, having a table of standard information (in relevant languages) is advantageous so that individual MSDS lists for each product can be created by selecting a combination of the standard information. Thus, changes required to the information need be made only once in the standard list instead of having to be made in every MSDS referring to the item in question. There will always be a need for free-form comments to be added to an individual MSDS in addition to the standard information.

An MSDS system will also track when MSDSs have been sent to specific customers. Various regulatory bodies have detailed procedures as to when MSDSs must be provided to customers and to employees. Often, sending an MSDS every time is easier than keeping track of when they are sent and when they are needed in the future. Nevertheless, a thorough MSDS system will keep this information and will provide the MSDS at the appropriate time instead of just flooding customers with paperwork. In addition, some hazardous materials require special labels to accompany the MSDS itself. These labels are attached to the packaging containing the product and/or the truck carrying the material. People dispatching the product must be notified of which MSDSs and labels are required under different circumstances.

A company with an extensive need for thorough MSDS control must integrate the MSDS system with the order-shipment process. The system reads the shipment information, interrogates the MSDS data base, and determines the MSDS information required for each shipment. At that point it can print either a MSDS for each shipment or a picklist of MSDS-related items that informs the dispatch personnel of requirements.

Sales and Bookings Analysis

All companies must maintain accurate and useful marketing information, the most basic of which is sales and bookings analysis. A "booking" is an order from a customer; a "sale" is shipment and invoice to a customer. The difference between the two represents items ordered but not shipped to the customer for one reason or another. This information is readily available within the order-entry and shipment part of the computer system and can be fed into history files for use in analysis and reporting.

Keeping detailed information about each individual order and customer shipment is usually not necessary. Stripping off the information from the sales order and shipment files into history files prior to deleting the details from the system is more advantageous. As the speed and capacity of computer hardware is increasing every day, there is less need to delete extraneous data from system files than in the past. However, it is still generally accepted that maintaining good housekeeping on computer files is better than letting them grow unchecked. The smaller and cleaner the files are, the more efficiently programs will run — and efficiently running programs result in fast, responsive systems.

Sales-and-bookings-analysis information can be summarized at the time the information is stripped from order and shipping information. Most marketing analysis systems allow the user to define the level of summarization required on sales and bookings files. Good sales-and-bookings-analysis systems allow for flexible reporting. The need for marketing information changes as the market changes, as the products change, and as the company changes. Marketing information must be accessed easily and relevant information provided to the people using the system. Raw data in the system must be maintained in straightforward files so that information can be accessed using flexible reporting tools.

Integration with Accounts Receivable

The order-entry and shipping system should be integrated with the accounts-receivable (AR) system. Entering information twice into computer systems is wasteful for two reasons. First, the work involved with entering the information twice over is wasteful in itself. Second and more important, having two separate systems containing the same information is wasteful, because the people responsible for each system must be sure that the two are kept precisely in line with each other. At best, this is a time-consuming process; at worst, it is the subject of fruitless strife and controversy within the organization. An important aspect of simplicity is that the information is maintained just once and with a minimum of transactions. The company is in the business of manufacturing and marketing products, not of making transactions into computer systems. Such extraneous activities are wasteful and should be gradually eliminated.

The integration of AR and order-entry systems means that customer information need be maintained only once, that invoice information has a single source, and that invoices are created in a timely manner. Many companies quickly recoup the cost of implementing an integrated AR system by the improved cash flow of sending out invoices as soon as the product is shipped. Also, activities like credit checking can be achieved on-line while the orders are being entered. Delays involved with passing an order from one department to another for checking this and verifying that can be eliminated — and with them a kind of waste that cannot be tolerated in an agile manufacturing company.

Summary

Many of the complex features of an order-entry and order-fulfillment process within a traditional manufacturing company are wasteful, either because they are poorly organized and executed or because they are used to control and monitor the delays, wastes, and inefficiencies within the organization. As a company moves into agile manufacturing, the computer systems — like the company itself — must be customer oriented, respond to customer needs, and provide flexibility. The goal is JIT delivery of make-to-order products with no delays and virtually no administrative procedures.

Systems are integrated to avoid wasteful repetition and to ensure a fast and complete analysis of product availability, pricing, credit control, and shipment. The systems are simple, because simplicity is essential to a world-class company, and yet focused to the needs of the customer and the company. The order-entry process varies according to the kind of products manufactured and according to the marketing strategy of the company.

The complex, multi-step allocation, picking, staging, packing, and dispatch processes common in traditional companies are simplified and largely eliminated. As the company moves toward zero inventory, short cycle times, and making to order, traditional, complex systems are systematically eliminated. Agile companies take advantage of new technologies like EDI to eliminate transactions and to reduce errors.

CHAPTER SIX

Additional Features of Customer Service

M ost agile manufacturing requires the basic capabilities of the customer-service software discussed in Chapter Five. This chapter considers three advanced aspects of customer-service systems that will become commonplace over the next few years: order-entry configurators, electronic data interchange, and distribution resource planning.

Order-entry configurators have been available for many years within manufacturing software used by companies that design to order or configure to order. As more companies apply agile techniques, the need to provide customers with more flexibility, wider product variety, and short lead times often opens up the need for a fast and responsive configuration capability during the order-entry process.

Electronic data interchange (EDI) is a customer-driven requirement. More and more customers are requiring suppliers to receive and provide information electronically instead of having large numbers of paper documents printed and mailed back and forth. Electronic interchange is faster, accurate, and saves the time and cost of manual data entry. EDI has been used extensively in the automotive industry for some time and recently has become prevalent within large retail chains. In time, EDI will become the standard method of transferring information, orders, and cash.

Distribution resource planning (DRP) applies the concepts of manufacturing's material requirements planning (MRP) to the distribution network. The complexity of a distribution network and its impact on customer service and cost has led many larger corporations to develop and use computer systems that keep track of which products are where and of how best to supply customers. While these kinds of systems are not new, they are not widely used among Western manufacturers. As the need arises for more responsive customer service and as technologies become less expensive and more available, companies that make to stock will be turning to DRP to maintain world-class customer service.

Order-Entry Configuration

Traditional order-entry systems, and indeed the MRP II systems that back them up, are very much oriented toward a make-to-stock environment. The system must define all products ahead of time, assign a product number, set up the appropriate part-master and warehouse-item records, create bills of materials (BOMs) and routings, establish product costing information, and set prices and discounts. When these processes are completed, customer-service people can begin to take orders from customers. A company moving into an agile way of thinking about customer service will begin to emphasize flexibility and responsiveness to customer needs, and these traditional approaches can become burdensome.

For many years companies that make to order have found traditional order-entry and production-planning systems unsuitable. In addition, companies that provide a wide range of options in the way the final product is configured find traditional systems cumbersome because each unique configuration must be set up on the system before an order can be placed and processed. Traditional systems offer two ways to overcome this.

One way is for engineers to attempt to provide every possible combination and configuration of the product and create the relevant inventory, pricing, and manufacturing records on the system. A little knowledge of the mathematics of permutations and combinations quickly reveals that even a small number of interrelated options can result in a very large number of possible finished products. Creating this "number" system becomes a huge task and results in enormous computer files containing hundreds (or thousands) of products that are rarely or never used. Even when every product is set up on the system, order-entry people have difficulty determining exactly which product number the customer requires. They must "translate" what the customer is requesting into the appropriate part number on the system.

Another solution to the problem is to create generic item numbers for the product family and enter the specific configuration in the form of comments when the customer order is set up in the system. While solving the problem of having to set up a large number of products, this method creates an additional problem because of the mistakes and misunderstandings that stem from the descriptions being free-form and imprecise. Either the wrong product gets made or there is an increase in production time because inquiries must be made to clarify the order. Because generic product-family numbers are used, maintaining useful product sales history is difficult, and the use of generic costing and pricing with a high degree of manual override becomes necessary.

A further complexity arises from this issue. Under these circumstances, customer-service people need a high degree of product knowledge to ensure that the correct order is entered, particularly when product engineering is such that only certain option combinations are feasible. For example, a car may require a stick shift when the high-powered sports motor is chosen. A bicycle company may offer certain features on mountain bikes that are unavailable on a road racer. These needs require either that the order-entry personnel be product specialists or that the process be lengthened by the need for a technical review of each order before its release to production.

The complexity, long lead times, and likelihood of errors inherent within these approaches to the make-to-order environment runs contrary to agile aims. A world-class manufacturer strives for reduced lead times, a high degree of flexibility, and no errors. In fact, many advanced manufacturers are working hard to move from making to stock to making to order. They are also providing customers with more options in order to be more flexible and responsive to customer needs. An order-entry configurator is one method that can provide greater flexibility and choice without the additional complexity and the resultant errors and delays that go with it.

What Is a Configurator?

A configurator is a set of functions added to an order-entry system that allows finished products to be configured as a part of the order-entry process. The three aspects to an order-entry configurator are:

1. a menu of available options to select
2. the inclusion and exclusion logic for validating selected options
3. the creation of manufacturing production orders

When an order line is entered for an item that is flagged as a configured product, the order-entry system invokes the configurator logic. A series of configurator menu screens presents available options to the user who can select the option required. For example, a motorcycle configuration may offer three engine sizes, five different frames, and seven accessory packs. The order-entry person selects the options required by the customer from the configurator menus.

The price of the configured item is calculated as the user enters the information and selects the options. There are a variety of ways to price configured products, and the methods included in the configurator will be in line with the company's approach to configuration pricing. One approach sets a price for each item and the configured price is the sum of these individual prices. Another approach is to have a standard uplift amount for the product; the configurator calculates the total cost of the configured product and adds in the uplift amount to derive the price. Some companies have a set price for the generic item irrespective of the detailed configuration.

Another more complex approach includes the production-routing information. As the configuration is built up, the configurator accesses not only component parts but also the routing required to manufacture the final assembly. This routing will add the associated labor and overhead costs into the total cost of the product before an uplift is applied. The uplift may be either an amount or a percentage. Some companies have uplift prices that are dependent upon the number or type of components selected; they use an algorithm to calculate the price from this information.

Frequently, the configurator screen will show the cost, price, and margin information so that the order-entry person can make sure the company is maintaining correct margins on the item. This person would have the authority to override a price based upon this information.

Configurator Logic

When designing a configurator, do not make the logic overly complex. A configurator that is difficult to use can become self-defeating.

The same care must be taken when setting up the configurator bills and the inclusion/exclusion logic. It is easy to make the system so confusing that people find it hard to understand.

Inclusion and Exclusion Logic

The feature of inclusion and exclusion logic validates the selected options. In the example of a motorcycle, it may be that the smallest engine size is compatible with only two of the available frames; "exclusion logic" would prevent an invalid combination being selected. Similarly, selecting one frame size may require selecting a specific accessory pack; "inclusion logic" would pick up this requirement and ensure that compatible components were selected. This logic would also know how many accessory packs could be selected at the same time and which of the packs are compatible with each other. The inclusion/exclusion logic reviews each selection made by the user and communicates both invalid selections and required selections as the configuration process continues.

Inclusion/exclusion logic can be quite complex. A typical configurator will use the logic each time a component is selected, and then do a final pass after the entire configuration is complete. There are extremes on this approach. Some configurators are very rigid and force the user to adhere to the predetermined logic. Others allow entry of any configuration, review the selection after the configuration is complete, and then provide error messages. The rigid approach is useful when the order-entry person has no knowledge or latitude to make changes and where the inclusion/exclusion logic is well formulated. The other approach is useful when the person entering the order has knowledge of the product as well as the authority to override the prescribed logic.

Entry of the menu selections and inclusion/exclusion logic is crucial to the success of a configurator. The more accurate and thorough the menu selections and the validation logic, the better will be the resultant configurations. Most configurators require an engineer or product specialist to carefully construct the option selections and to precisely define the inclusion/exclusion logic. The logic often consists of a series of "truth tables" that define, for example, that option X does not go with option Y, and if option Y is selected, then options W and Z must be present. However, a minimum of three and maximum of five options can be selected and these tables can become very complex.

Some of the more advanced configurators apply artificial intelligence to the configuration logic. These kinds of systems "learn" as they go along. Instead of all the validation rules and combinations being entered up front, an intelligent configurator allows the engineer or product specialist to create valid combinations, building up its own set of rules based upon the "experience" it gains from use. These so-called expert systems are in their infancy but can potentially become very powerful in this area.

This approach will often be able to validate a configuration more quickly because of an heuristic approach to the analysis. The rule-based approach requires that all the truth tables are read and applied every time a configuration option is selected. An heuristic approach takes a guess at the analysis and then validates the guesses it has made. This method of analysis — which approaches a problem the same way that people analyze problems — can often respond quicker than a rule-based system. Part of the attraction of expert systems is that they learn from experience in a way that is similar to human beings. This capability makes them less intimidating and more readily accepted by the people using them than a system with rigid logic.

Creation of Manufacturing Production Orders

Once an order line has been configured, a manufacturing production order can be initiated. The configuration process creates a bill of materials of the major components that make up the finished product. This bill of materials can be moved into a work order or production schedule so that the product can be made. The configured product is usually given a temporary product number, which can be derived from an odometer or from the sales-order number and line-item number. Once the product is manufactured and shipped to the customer, the temporary product number is removed from the file. The bill of material will have the same temporary number and be held on a configurator file that can be used as a permanent record of the specific configuration ordered by the customer on this occasion. The configuration must be retained for traceability purposes and so that a similar product can be located for future production.

Typically, a configurator will pick up a standard production routing from the generic product record that is carried over into the production order. This routing may be as simple as defining the production cell in which the item is manufactured — or it may be more complex. Some companies require the configurator to create a routing based upon the selections chosen for the configuration. This system requires a specific algorithm or routing selection method that analyzes the configuration and determines which activities are required to complete the product. Some manufacturing and distribution systems do not require a manufacturing production order to be created when initiating production. They allow sales orders to drive the production plant and the sales order itself becomes the authority to manufacture the item. Creating a new set of documents with a new number is unnecessary because all transactions are recorded directly against the sales order itself. This approach is simpler and more direct.

Many agile companies offering short lead times and fast turnaround on configured items do not require any kind of production order at all. The sales order creates a printed configuration that is used by final assembly to make the item without any work orders or routings. No transaction, inventory, or timing information is entered into a scheduling system; the entire production is completed manually from the sales order itself. The simplification that comes from short cycle times and just-in-time (JIT) production allows for a more simplified final assembly of configured items.

Electronic Data Interchange

One of the most important and fastest growing aspects of computer integration is the use of electronic data interchange. EDI can be used for receiving customer orders, placing purchase orders, transmitting documents (e.g., shipping notices, invoices, quality information, specifications) and for transferring money. All of these transactions can be achieved by electronically transmitting the information in a standardized form from one trading partner to another. EDI can eliminate printing and mailing paper documents with their associated costs, delays, and errors.

EDI has been available for many years and some companies have been using it successfully, particularly for internal transactions. Not until the recent establishment of

---

Trading Partners

The term "trading partner" is used within EDI to denote any relationship between two companies or entities. Customers and vendors are trading partners, and almost all EDI transactions are information transfers between a vendor and a customer. A standard EDI transaction is called a purchase order. This same transaction is a purchase order to the customer and a sales order to the vendor. The terminology can be confusing; hence, the choice of words.

---

standard formats and methods has EDI become a practical option for most companies. EDI will soon become the standard method of communicating between trading partners. *The New York Times* noted that "It would be almost unthinkable nowadays for a business not to have a telephone to communicate with customers and suppliers. In the future, it may be almost unthinkable for a business not to have a computer for the same purpose."[1] *The 21st Century Manufacturing Strategy Report* lays out an industrial future in which there is vast and rapid movement of information worldwide. Electronic communication becomes so effective that individual companies grow less significant because partnerships and liaisons can be easily created, used, and abandoned according to need.[2]

Reasons to Implement EDI

There are a number of reasons for using EDI. Professor Margaret A. Emmelhainz, a specialist in EDI technology, has identified six principal ones: business survival, cost savings, improved operations, improved customer responsiveness, improved channel relationships, and improved ability to compete globally.[3] These are all objectives of agile manufacturing. In brief:

- EDI facilitates the fast, effective transfer of information between trading partners.
- EDI significantly reduces transaction costs.
- EDI makes frequent (daily or more often) placement of JIT delivery orders practical.
- EDI improves the accuracy of information flow.

However, in reality most companies introduce EDI because their customers require it. Automotive manufacturers require EDI placement of orders with their first-tier suppliers, particularly for JIT deliveries based on production schedules. Some automotive companies have made EDI an aspect of their vendor certification programs — vendors must accept and provide EDI transactions if they wish to do business in the longer term. Many automotive suppliers are passing down these same requirements to their vendors.

Major retail chains and mail-order houses in the United States and Europe now require EDI transactions of their suppliers. Sears Roebuck requests EDI order placement for both

---

EDI and Agile Manufacturing

EDI incorporates several key aspects of agile manufacturing. It provides fast and accurate information flow, reducing waste. It is simple to use, once the initial technology has been mastered, and provides a standardized process of information transfer.

EDI incorporates the concept of partnership between customers and vendors. It provides a method for the mutually advantageous transfer of information. Although just a piece of technology, EDI has the potential to be a significant vehicle for introducing WCM methods.

its retail stores and its all but discontinued mail-order operations. Sears mail-order operations extended the use of EDI so that, instead of the supplier shipping product to a Sears warehouse, the end customer's order was electronically passed to the supplier, who shipped the product directly to the customer. This approach had the potential to eliminate inventory for Sears and to improve the level of customer service.

Of course, approaches like this place additional burdens on suppliers. Some companies feel coerced by the "big boys" into assuming more risks and higher costs without receiving adequate compensation. World-class companies will work hand in hand with their trading partners to create a "win-win" arrangement that is mutually beneficial over the long term.

How Does EDI Work?

Electronic data interchange is applicable to a wide range of business activities, but this discussion concentrates on the use of EDI for customer order entry.

The purpose of EDI is to move information from the customer's purchasing system into the vendor's order-entry system. Transfer of information is facilitated by using standard format and record layouts. When the customer has approved a series of purchase orders, the information is accessed from the purchasing system and converted into a standard format for transmission to the vendor. The file is transmitted to the vendor. The vendor's system receives the order information in standard format and translates it into a format that can be read into the system and can create a sales order. Information transmitted using EDI in a standard format contains all data necessary to complete the transaction. These data include item number, quantity, required delivery date, delivery information, price, and terms.

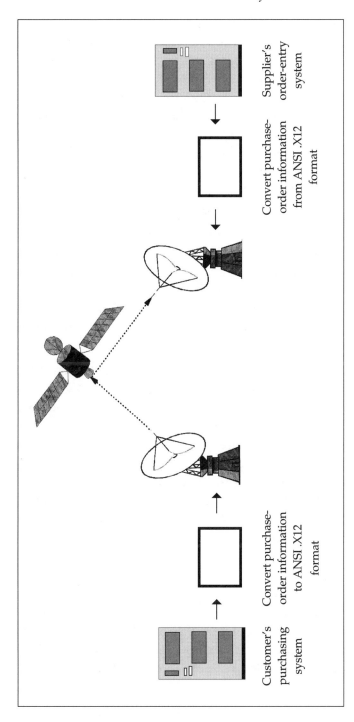

Figure 6-1. Using the EDI Network to Transmit Purchase Orders

Frequently, the file is created from reading recommended orders from the customer's MRP or procurement system. This file is read into a program that "maps" the information into a file in one of the standard formats (X12 in the United States, EDIFACT in Europe) ready for transmission. The file may be transmitted directly to the vendor, particularly if the vendor is another part of the same organization. In most cases, information is exchanged using an EDI network that is often called a value-added network (VAN).

A VAN operates in the EDI world the way a post office delivers letters. A trading partner places all the EDI orders into its VAN "mailbox." The VAN sorts the orders and "delivers" them to the vendor's mailbox. The vendor can then retrieve the orders from the mailbox at a convenient time. Consequently, any direct connection between the trading partners is unnecessary because the VAN handles any transfer of EDI transactions.

There are a number of these networks within the United States and Europe and worldwide. They not only transfer information but also track transactions, provide backup and security services, and give users detailed information about their network usage. VANs remove much of the complexity from the EDI operation. Some major corporations have their own network for the exclusive use of their trading partners. These networks commonly allocate individual time slots within which each trading partner can transmit and access transactions within the network.

EDI Transmission Standards

The introduction of national and international standards for the electronic transfer of information has been key to the widespread acceptance of EDI. The American National Standards Institute established the ANSI X12 standard during the 1980s in association with numerous technical and professional organizations, including the Automated Clearinghouse Association, the National Association of Purchasing Managers, and the Automotive Industry Action Group (AIAG). ANSI X12 defines the file-record layouts used for transmitting information related to a range of specific documents. More than 200 documents are currently included in the X12 specifications and this number is increasing as EDI is extended into more and more areas of business and government. Some documents of interest to manufacturers and distributors include the following:

- X12.1   850   Purchase Order
- X12.2   810   Invoice
- X12.7   840   Request for Quotation
- X12.8   843   Response to RFQ
- X12.9   855   Purchase-Order Acknowledgement
- X12.10   856   Advanced Shipping Notice
- X12.12   861   Receipt Advice
- X12.14   830   Planning Schedule
- X12.15   860   PO Change Notice

The European standard, quickly becoming accepted as the international standard, is called EDIFACT. This system is sponsored by the British Standards Institute and the related standards organizations within the European Economic Community and other European countries. ANSI has introduced a more comprehensive standard called ANSI X400, which combines both EDI and broader electronic-mail ("E-mail") standards. This system facilitates the standard interface of documents, free-form descriptive text, and graphic images. The intention is that all aspects of electronic communication will be standardized through X400 including EDI, E-mail, fax transmissions, and the burgeoning technologies related to computer imaging. However, most manufacturing and distribution organizations find X12 or EDIFACT to be sufficient for current needs.

Software Needs of EDI

A number of software packages have been developed to meet the needs of EDI. Many of them, written for personal computers and usable on a stand-alone basis, allow X12 or EDIFACT data to be transmitted and received. The intention is to interface these systems with a company's order-entry, production-planning, and procurement systems. However, many companies use these systems merely to receive and transmit information. Data are keyed manually into and out of the EDI system, serving the customer's requirement for EDI interaction without providing any EDI advantages to the user company.

A fully functioning EDI system is integrated (or at least interfaced) with the company's other systems including order processing, procurement, and financial applications. The EDI system serves three primary functions:

1. It interacts with the network.
2. It maps data to and from EDI.
3. It interfaces with other systems.

Protocols and procedures are required to interact with the standard EDI networks, including definition of trading partners, scheduling of data transfer, and monitoring the use of the networks. These protocols are quite specialized and require software that specifically addresses these needs and is kept up-to-date with the changing criteria of the network companies and major trading partners. For example, some large corporations have very large amounts of EDI data to transfer and require trading partners to interrogate the network at specific times during the day. The EDI software will ensure that the network is accessed during the available "time window" and will record the successful transfer of data. One vendor to General Motors has an early-morning time slot. The EDI software has been set up to access GM's network at the right time in the morning, monitor the transmission, monitor the company's two minicomputers, automatically switch the received data between the two machines, and telephone the system manager at home and wake him up if there is a problem the software cannot fix.

The process of mapping data, an important aspect of an EDI system, translates data from the standard format to or from a format that is acceptable to the company's other systems. For example, the X12 purchase-order information (type 850) may contain data fields that have no equivalence within the company's systems, or the field may be formatted differently. The EDI software will have facilities to load default values to essential fields, to adapt field formats, and to translate data into or from the standard format. The better EDI packages have an easy-to-use mapping procedure that allows the changing needs for data interchange to be easily met without the need to write new programs each time. These mapping processes enable users to provide the information to map one field to another using a simple matrix that the system then uses when translating the data.

Integration between the user's business software and the EDI system is at the heart of the interface. These programs can be quite simple when data is fed out of the systems into the EDI system (i.e., placing purchase orders). Relevant information is merely stripped from the standard system in a format the EDI system can map into a standard format. Additional fields are needed on some files to tell the system which transactions are being processed using EDI, who the trading partner is, and the status of the transmission.

Data entry into business systems using EDI can be much more complex. In reality, the process is a batch update

---

EDI Software

Do not skimp on EDI software. The simple, inexpensive PC packages currently available often lack the broad functionality and flexibility of larger, more comprehensive packages. In general, having your EDI software run on the same machine (or network) as your business software simplifies data transfer and provides a more integrated approach.

Make sure the EDI software has an easy-to-use data-mapping feature. EDI standards are growing and changing all the time. Having the ability to map your own records is important.

---

of the system (i.e., batch order entry). The EDI entry program must fully emulate the on-line processes and report any errors or warning messages. When EDI entry programs are tailored for a specific company's needs, the features and functions may be less complex. However, standard packaged software must be able to process every entry, change, delete, and completion feature of the manual on-line entry system.

The use of shared routines can be a valuable part of the structured programming design of on-line and EDI entry programs. Much of the coding can be shared between the programs and will make enhancement and error correction easier. Setting up such a system is not easy, however, because characteristics of an on-line system and a batch system are very different. The on-line system needs to perform validations and report errors immediately on the screen, whereas a batch update performs all validations at an early stage and rejects invalid entries, writing them onto an edit report.

Required EDI Transactions

Not all EDI transactions have to be available in a firm's business systems before the benefits of EDI can be reaped. Using EDI effectively to receive orders from customers and to process those orders requires having the purchase-order entry (850), the purchase-order change (860), the advanced shipping notices (856), and the invoice (810). This basic capability enables interaction with a customer and provides customers with the information they need.

Using EDI to place orders on vendors requires the same transactions. The 850 is simply used by stripping purchase-order information from the purchasing system and by formatting for EDI transmission. Processing an advanced shipping notice, which is a more complex procedure, can be used to emulate a goods-received transaction or to create an in-transit record within the system. This is then updated when the goods physically arrive. Similarly, invoice entry through EDI can be quite complex. The program must emulate the accounts-payable invoice-entry routines within the system and create the appropriate records and update files the same way the manual entry of invoices would.

Including an invoice and receipt matching process for EDI orders is often unnecessary, because there are far fewer errors to detect. Receiving and invoicing information is derived from the same place and is transmitted — often simultaneously — through the same process. Most invoice and receipt matching errors are caused by human error, which is largely eliminated by the use of EDI.

Companies that offer engineered-to-order products can use the EDI transmission of specifications and bills of material as a key competitive feature. Receiving engineering and design information through EDI can provide enormous savings under these circumstances. This trend is growing and is part of the movement toward flexibility, allowing companies to receive specifications directly and process orders onto the shop floor far more quickly and accurately.

Electronic funds transfer, a branch of EDI that electronically handles the payment and receipt of cash, is used extensively within Europe but is less popular within the United States owing to the more fragmented banking systems. Electronic funds transfer is another growing trend. If invoices can be transmitted electronically, then cash can be transmitted just as easily and quickly. Clearly, there must be careful control when cash is moved electronically — however, no more so than when orders are placed and shipped electronically. The accounting or treasury departments of many companies are somewhat reluctant to use this method for the payment of invoices. Two areas where EDI is used extensively for the movement of cash are (1) for automatically depositing paychecks into employee bank accounts and (2) for retrieving lockbox deposits.

Distribution Resource Planning

While distribution resource planning is considered a new technique by the majority of manufacturing and distribution companies, it has been used for many years in some organizations. (Abbott Laboratories in Canada pioneered DRP techniques in the late 1970s.[4] And in 1981 the author worked on the team that designed and implemented a worldwide DRP program within the Xerox Corporation.[5]) Only in recent years has DRP come to be seen as an essential aspect of inventory logistics.

DRP is simply the application of time-phased requirements planning to a distribution network. In the same way that MRP uses time-phased requirements planning to schedule raw materials and components for production, so DRP uses the same techniques to schedule the replenishment of warehouses and depots throughout a company's logistic chain. DRP is used by companies that predominantly make to stock either their primary products or the spare parts and accessories that support these products. These make-to-stock items are warehoused throughout the country or the world to support the needs of local customers and to provide fast delivery. The objective of these distribution operations is to provide the customers with the best possible service at the lowest cost.

Agile manufacturing applies the same approach to the distribution operation as to the manufacturing operation. The ideas of JIT, total quality, and employee involvement are just as critical to the logistics organization as they are to the factory. The reduction of inventory, cycle times and customer lead times, and waste have radically changed the way world-class companies view their warehousing processes. Some companies have eliminated the need for warehousing by building production plants adjacent to the plants of their major customers. Other companies, like John Deere Corporation, have opted for having many smaller plants throughout the country instead of having one or two huge production facilities. Agile manufacturing concepts have superseded the old ideas of economies of scale and mass production, which tended toward centralization. Nonetheless, there is still a need for many companies to have a distribution network for their products and spare parts.

Traditional companies invariably use inventory-planning systems in their warehouses that make use of reorder levels (ROL) and reorder quantities (ROQ). An ROL for each item defines the minimum amount that must be in stock before a replenishment order is placed on the supplier or manufacturing plant. When inventory falls below this level, an order is placed. The ROQ is the amount of that item ordered each time. A number of techniques are used to calculate the ROL and ROQ quantities. Some methods are very sophisticated and use statistical analysis to determine ideal quantities and to match these quantities to the desired customer-service level. Other methods are more pragmatic and are based on the number of weeks of stock the company wishes to keep on hand. The level of complexity and sophistication increases when the company makes items that are highly seasonal or has many promotional campaigns.

These traditional methods of inventory planning have the following shortcomings:

- When each warehouse is controlled independently, one will have excess stock while others have shortages.
- There is no advanced warning to the supplier or production plant of when replenishment orders will be placed by a warehouse.
- The ROL/ROQ approach amplifies small changes in customer demand into large demand fluctuations on the supplier or production plant.
- Production plants have no way of knowing the total demand of all warehouses. Most plants plan production on the basis of sales forecasts and erroneously fail to take account of replenishment methods used within the warehouses.
- ROL/ROQ is inflexible to changes in demand. Increased demand invalidates both, resulting in shortage and expediting. Reduced demand creates excess inventory because there is no mechanism for cancelling replenishment orders or for updating the ROQ and ROL when demand changes.

DRP overcomes many of these problems and provides a consistently higher level of service with less inventory. In his book *Focus Forecasting and DRP*, Bernard Smith identifies the following eighteen advantages of using DRP over traditional methods:[6]

1. DRP orders inventory.
2. DRP cancels and reduces excess inventory.
3. DRP tells factories and suppliers about future orders.
4. DRP allocates scarce inventory.
5. DRP identifies excess inventory.
6. DRP redistributes excess inventory.
7. DRP provides summary measures of performance.
8. DRP projects resource needs in value, weight, cube, hours, and so forth.
9. DRP uses exception reporting.
10. DRP does break bulk allocations.
11. DRP gets steady input from major customers.
12. DRP creates action messages.
13. DRP matches the needs of service turnover to capacity.
14. DRP converts customer item numbers to the manufacturer's.
15. DRP does joint replenishment for trucks, cars, and containers.
16. DRP handles multiplant, multiwarehouse inventory management.
17. DRP provides input for EDI.
18. DRP allows effective use of JIT logic.

How DRP Works

The mechanics of DRP are simple. The system knows the stock level at each warehouse for each item. It also knows the total forecast for each item at each warehouse. Each item must be forecast at the warehouse level, and special requirements like promotions, special customer orders, and interplant transfers must be added to the forecast. The system needs to know the safety stock for each item at each warehouse, and safety stock is usually expressed in terms of "safety time." (Safety time is the number of days or weeks of inventory that should be in stock at any one time.)

From this information the system can calculate the expected on-hand balance for an item over as many future periods as required. It can also identify when the projected on-hand balance falls below the safety stock and place a replenishment order for the appropriate quantity. This replenishment order can be rounded to a pack size or pallet quantity, but will not be based on any concept of economic order quantity. The goal is to replenish only the amount needed to serve the customer. An example is presented in Table 6-1.

DRP calculations are done for all items at each warehouse and consolidated orders placed on the manufacturing plant or supplier for each item. DRP orders show not only what is required immediately but also planned future requirements. This additional information allows the manufacturing plant to plan future production based upon reliable projections.

When manual intervention is required, the planner will be notified by the system to review an item. The system requires the planner to take action only on exceptions such as major changes in demand or changes to replenishment order quantities inside of the lead time. In this way, thousands of items can be controlled effectively without the need for manual checking or validation of every item; only the problem parts are highlighted for review.

Logic can be built into the DRP calculations to redistribute excess inventory from one warehouse to another. Generally, redistribution should be done as a separate exercise rather than as part of regular, ongoing inventory-control procedures. The redistribution logic will create transfer orders authorizing movement of materials from one warehouse to another.

122S1096 1/4" Brass bolt	Mar 3	Mar 10	Mar 17	Mar 24	Mar 31	Apr 7	Apr 14
Total requirements (forecast + specials)	0	58	55	42	35	40	45
Scheduled receipts				100			100
Expected on-hand	102	44	-11	47	12	-28	27
Safety stock (1.5 weeks)		86	76	60	55	63	72
Planned orders (rounded to pack size 10)		50	40	30	30	50	50
Planned on-hand		94	79	67	62	72	77

Table 6-1. Distribution Resource Planning of a 1/4" Brass Bolt

Transportation Logistics

When a company has a significant distribution network, the cost and effectiveness of the transportation logistics are important. There are two primary issues — the long term and the short term. The long-term issue relates to capital investment in new warehouses; that is, when they are needed and where they should be located. The short-term issue relates to how many trucks, rail cars, or containers are required today or this week to move material from one location to another. Both of these issues take on still more importance as a company moves into JIT deliveries of smaller quantities on a daily instead of weekly or monthly basis.

Short-term issues of the availability of transportation and of optimizing cost by consolidating loads, ensuring return loads, using common carriers, "milk rounds," and so forth are assisted greatly by information made available through the DRP system. DRP provides detailed information on weight, cube size, number of boxes and pallets, and value of the material to be transported. This information is available not only for the immediate period but also for future periods.

Certain specialist transportation systems can take information from the DRP system and create the shipping schedules and documentation required to optimize the movement of the products. Owing to the nature of the transportation business, there are few generic transportation optimization systems. Most systems tend to be customized specifically for the needs of the organization using the system, and they frequently use quite complex and sophisticated mathematical analysis to determine the most advantageous way to move goods. These methods include both heuristic optimization methods and regression analysis to solve the problem of optimizing across many conflicting variables.

Some major transportation companies provide services where the DRP information can be fed into the transportation company's analysis software, and the responsibility for transporting the goods in the most effective way lies with the transportation company. Some companies, like Ryder Corporation, have special transportation services for companies operating a JIT delivery service to customers. This subcontracting of transportation services has proved most effective for some agile manufacturers.

World-class companies often simplify their transportation processes by making close working arrangements with their major suppliers for regular JIT daily deliveries (or more often) of similar quantities. The objective is to remove much of the uncertainty and variation in the transportation process. Just as a manufacturer removes variation in the production process to ensure quality, so the distribution organization removes variation from the transportation process to achieve quality delivery. For a large organization, attaining this goal often requires developing a simple, yet sophisticated logistics network for delivering materials to customers and internal warehouses using full truckloads and consolidation, while eliminating wasteful delay and activities.

The key to the success of much of this activity is the availability of reliable future transportation requirements several days or weeks ahead of time. This information is available from a well-designed DRP system. Some companies try to forecast their transportation needs independently of inventory-control systems, usually because the systems are inadequate. A good DRP system provides the information needed to optimize short-term transportation requirements.

Long-Term Transportation Optimization

Questions relating to where and when to build a new warehouse or distribution center are long-term strategic issues. Software used to support these decisions are complex analytical tools that use regression analysis (or other sophisticated statistical analysis tools) to optimize transportation costs and customer-service levels based on projected activities over a long planning horizon. Ultimately, these decisions have more to do with management judgment than software-assisted analysis.

Agile manufacturers tend to view these issues in a simpler way than traditional companies. The optimum is to have many small and effective production plants adjacent to the company's customers. Next best is to have many distribution locations adjacent to the company's major customers. As with all other strategic decisions, the emphasis is on simplification, customer satisfaction, and high quality. Computerized analysis, although a helpful technique, is only one of many other considerations studied when making these decisions.

Summary

Three aspects of computer systems are increasingly prominent within agile manufacturing companies. They are order-entry configurator, electronic data interchange, and distribution resource planning. Previously considered advanced, these methods are being accepted as a standard part of world-class procedures.

As a company moves from a make-to-stock to a make-to-order environment and as the requirement for a wider variety of products becomes more prevalent, order-entry people must have tools that allow them to configure the order according to customer needs instead of placing orders for standard stock items. An order-entry configurator is the part of the order-entry process that provides a systematic way of configuring nonstandard products. The attributes of the product are selected from features and options that have been previously set up within the configurator. The inclusion and exclusion logic within the configurator ensures the validity of the selected combination of features and options. The configurator optionally creates a production order using the select-ed features and options as the bill of material required to manufacture or assemble the final product.

Electronic data interchange is a standard approach to electronic communication between companies or trading partners. The format and content of the data records used to transmit information between companies have been agreed upon by various standards bodies throughout the world. Agile manufacturing uses EDI to provide fast, flexible, accu-rate information exchange with vendors and customers. An additional benefit is that there is no manual intervention in the process of creating orders, sending acknowledgments, reporting shipments, sending invoices, engineering data, and other EDI transactions. Replacing wasteful manual transac-tions by automatic electronic transaction enhances accuracy and reduces waste.

Distribution resources planning is required by compa-nies having a multisite distribution network. DRP applies MRP features of time-phased orders and netting require-ments against supply across a multiple network of warehouse locations. The objective is to source the supply of a product to meet customer orders in the most effective way. This objective requires balancing the customer's need for short lead times with the company's objectives of zero inventories and low cost. DRP assesses the optimum supplying location to meet customer requirements and organizes the replenishment and redistribution of materials to best fulfill the objectives of agile distribution needs.

CHAPTER SEVEN

Procurement and Vendor Relationships

*R*adical changes occur in the area of material procurement when a company moves into agile manufacturing. Traditional ideas are turned on their heads by new concepts. The average purchasing professional, whose entire education and career has been built upon those traditional ideas, has great difficulty accepting these changes. Implementing a new approach to procurement is a major challenge to a world-class company.

With regard to procurement, two principal mistakes made by companies moving into agile manufacturing are to give it either too high or too low a priority. Companies that give just-in-time (JIT) procurement too high a priority expect vendors to provide JIT deliveries, zero defects, and price reductions immediately, when they themselves have not yet implemented world-class techniques. You cannot expect suppliers to do what you do not do yourself; in other words, get your own house in order before making extraordinary demands. Many large, influential corporations having considerable power over their suppliers gave ultimatums to provide JIT deliveries or cease being a supplier. Suppliers were forced into JIT deliveries and often resorted to holding large finished-goods inventories that benefitted no one in the long term.

Many companies, finding the new procurement concepts too radical for their purchasing people to accept, have not attempted to make the changes required. These are the companies that mistakenly do not emphasize procurement changes enough. If the company is to be successful, every aspect of agile manufacturing must be implemented because effective implementation by half measures is impossible. The importance of each aspect and the emphasis placed upon it will vary considerably from one company to the next, but all aspects must be addressed.

While it is true that changes in procurement methods should often be addressed later in the implementation process, they do need to be addressed fully and radically. The changes are fundamental and represent a different philosophy of the company vendor. Such changes take a long time to develop, and there must be active change as a part of the move toward world class manufacturing. Ignoring this aspect is a serious error.

Just-in-Time Delivery

The first requirement of a supplier in an agile work environment is the ability to deliver just-in-time. In traditional companies the vendor will deliver products weekly or monthly, sometimes less often, and the company will hold raw material and component inventory until production needs it. The total quantity required is calculated using MRP or some other method of material planning and then rounded up to economic order quantities (EOQs), truckloads, or pallet sizes. MRP commonly sums up a week or a month of stock, so that the supplier makes few deliveries of large quantities. This approach is based on the EOQ concept, which is calculated by matching the costs of placing the order and receiving material with the cost of holding inventory. When this match is made, a quantity is calculated that will theoretically be the economically optimum quantity.

The concept of economic order quantities is unknown in an agile environment where the emphasis is on zero inventory. Zero inventory is best achieved by having the vendor deliver today what is going to be used today — no more and no less. This level cannot always be achieved for all components and materials, but the majority should be delivered directly to the shop floor on a daily or twice daily basis and used immediately for production.

Single Sourcing

Traditionally, multiple suppliers are utilized for each component, raw material, or service so that one vendor can be pitted against another and the lowest prices obtained. There is also safety in multiple sources because if one supplier fails, another can supply the material. Agile manufacturing takes the opposite approach by deliberately aiming for single sourcing of materials. The objective is, instead of spreading the business across many suppliers, to do a great deal of business with a few suppliers and to develop close working relationships with these suppliers.

Close and mutually beneficial relationships take time to develop and cannot be achieved with a large number of suppliers. It is necessary to limit the number of suppliers and to work together to the point where distinction between the two companies becomes almost blurred. There is communication at every level — not just the buyer addressing the sales-

---

Guidelines for Effective Partnerships

In his book *Purchasing in the 21st Century*, John Schorr laid out the following guidelines for effective partnerships with suppliers. [1]

1. Trust and communication must be open and honest.
2. Suppliers are part of your organization; they just happen to be outside your gate.
3. Shared risk means fair price and fair profit.
4. Long-term relationships are better than one-night stands.
5. Partnership consists of a great deal more than signing papers or issuing some edicts.
6. You cannot mass-produce partnership.
7. Performance is the main criteria for picking a supplier.
8. Internal accountability is the key factor.
9. Sell the concept to management; get the support.
10. Use an existing supplier to kick off the system.
11. Develop a working relationship.
12. Communicate constantly.

---

person. In both companies, production people speak to production people, quality people work together to ensure zero defects, design engineers work together in teams, and logistic personnel create effective transportation approaches. A level of trust develops so that all product and marketing plans and engineering, financial, and costing information is shared. Suppliers often take on responsibilities previously assumed by the design engineers. The thinking is that the supplier making the item is probably better able to design it for functionality, manufacturability, and low price because of its expertise in those aspects of the product.

Development of close working arrangements and mutually beneficial cooperation cannot take place with the adversarial approach traditional companies take toward their vendors. A philosophical change must take place.

Zero Defects and On-Time Deliveries

If a company is to work on a JIT basis, deliveries from vendors must be on-time every time. If deliveries are to be made directly to the shop floor without any receiving and inspection process, then the items must have zero defects. There can be no quality problems with materials or components. This level of excellence requires the vendor to be an agile manufacturer.

Developing the techniques and procedures to ensure on-time delivery and zero defects takes time, particularly if vendors are new to these ideas. Vendors that are already agile manufacturers themselves are increasingly using this competitive edge and guaranteeing on-time deliveries and zero defects. However, more traditional companies need additional time to improve their customer service. Close cooperation between vendor and customer will help overcome these problems and implement the necessary procedures.

Although there may be a period of change where the two companies work together to bring about improvement, the objective must be to make on-time deliveries and zero defects a reality as soon as possible. They are essential to JIT manufacturing. Additional requirements are short lead times and flexibility. Implementing these spin-off aspects of agile manufacturing at the same time is often possible.

While the cardinal need is on-time deliveries and zero defects, a world-class manufacturer will also look to reduce costs in the longer term. Here, a leap of faith is required because the implementation of agile methods will enable the supplier to become so much more productive that the real cost of materials and components will come down. The two companies must share cost information so that they can work together to eliminate waste and reduce cost. A world-class company will not force suppliers into financial jeopardy by imposing unrealistic price cuts — the long-term relationship and mutually beneficial partnership is more important than a few pennies off the price. However, lower prices are an inevitable longer-term result of the supplier's more agile production operations.

Vendor Certification

For a company to embark on a JIT manufacturing approach without the assurance of zero defects and on-time deliveries from its vendors would be irresponsible. In reality, developing the close working relationships with vendors takes time, and JIT deliveries must be introduced with careful planning. Many agile manufacturers continue to hold significant inventories of raw materials and component parts for some time after introducting JIT manufacturing methods because their suppliers do not yet provide the required level of service.

It is important to work with the vendors to create the methods necessary for zero defects and on-time deliveries. "Vendor certification" is a standard procedure for determining the ability of a vendor to provide the world-class service required by JIT manufacturers. It is also important to monitor the vendors to ensure that they achieve the agreed-upon goals.

In the vendor-certification process, a company creates the close-working relationships necessary for agile manufacturing. Agreement is made between customer and supplier on the timetable for reaching 100 percent on-time deliveries, zero defects, flexibility, and price reductions. This timetable may span several months to allow the vendor time to implement agile methods. As the plan is activated, the customer monitors the vendor to ensure that the agreed-upon improvements are, in fact, taking place.

There are levels of certification. The lowest level is awarded to a vendor that has a plan in place to achieve zero defects and on-time deliveries, although the customer continues to monitor deliveries and inspect materials. As the vendor's service improves, a higher level of certification is awarded which recognizes a higher level of service. This improvement allows the customer to cut back the amount of inspection and monitoring. The highest level of certification represents an understanding between customer and supplier that zero defects and 100 percent on-time deliveries are required and that the supplier consistently achieves this. At this point, the customer can accept direct deliveries to the shop floor without inspection or monitoring.

Software for Procurement

The problem requiring resolution is how to simplify the procurement, receiving, and accounts-payable processes. If vendors are to make deliveries every day or twice a day directly to the shop floor, how is that to be controlled? A traditional procurement system is replete with checks and validations to ensure correctness and accuracy and to prevent deliberate fraud. All these procedures represent waste that must be eliminated.

Traditional Procurement Process

A traditional purchasing-process system starts with a requisition that is either raised by an individual within the company or generated by the MRP application. These requisitions are printed and reviewed by a buyer or purchasing officer who determines the vendor and raises a purchase order.

In most systems the purchase order can be converted from the requisition without having to reenter all the information. The purchase order may have many lines, each representing a different part number supplied by the vendor. Each purchase-order line can have more than one delivery associated with it so that the purchase order provides the vendor with all the information required to schedule delivery. The vendor will then enter all this information into an order-entry system so that the materials can be manufactured and the deliveries made.

The next stage, the receiving process, involves the physical receipt of the product and processing of a receiving transaction. The receiving transaction matches the actual receipt quantities against the quantities ordered on the purchase order. It often includes specific functionality for spreading the receipt quantity across several delivery schedules to accommodate difficulties of late or early receipt of materials. Once the receipt quantity has been validated and spread across eligible delivery records in the purchase order, the system updates the purchase-order information to record the receipt and creates an inventory transaction to update stock records.

If the items received require inspection, then the system uses a "dock-to-stock" procedure whereby the item is shown in inventory but is not yet available for use by the shop floor because it awaits inspection and placement into raw-material inventory. The dock-to-stock procedure allows the goods-receiving personnel to record movement of the material from the receiving dock into the inspection process and then into raw-material inventories. If the item is rejected for any reason, the dock-to-stock process will track it into the material-review-board process whereby an authorized inspector or engineer will decide if the material should be returned to the vendor, reworked, or scrapped. This dispositioning process has several transactions associated with it — including scrapping, raising a return-to-vendor document, or processing a rework procedure.

Financial transactions also start at the point of receipt. The material can be received at standard cost, in which case the inventory value is debited by the amount of the receipt. If the company uses an average cost method of inventory valuation, then the new average is calculated from the current average cost and the actual cost of the newly received items noted on the purchase order. Some systems record a purchase-price variance (PPV) at this stage. (PPV is the difference between the standard cost and the purchase-order cost.) It is also possible to override the price at the time of receipt and perform all the calculations with the overridden price. This approach is commonly used only when receiving market-sensitive commodities where the price varies from day to day. According to the accounting rules of the company, the PPV may or may not be posted to the general ledger at the time of the product's receipt. Many companies wait until the invoice is received from the vendor.

When the invoice is received, a three-way matching is performed. In this process, the accounts-payable personnel enter detailed receipt information from the invoice. The system then matches the invoice to the receiving information to ensure that the invoice is in line with the quantity actually received. Providing there has not been too much delay between the time the goods were received and the invoice entry, discrepancies can be resolved by liaison between the accounts-payable personnel and the goods-receiving people. Unfortunately, there is often considerable delay between the two events and the reconciliation is difficult to perform.

Procurement and Agile Manufacturing

The agile manufacturer wants to eliminate this wasteful, complex, and error-prone process. The computer system must also have the capability to simplify the purchasing, receiving, and inspection process.

The ultimate objective is to eliminate the entire process because nothing about it is value-added. In reality, dismantling these procedures takes considerable time and is closely linked to the supplier certification process. Need for daily or twice daily deliveries puts an enormous strain on a traditional procurement approach. There just is not enough time to process all the transactions. For example, if a vendor makes daily instead of monthly deliveries, the number of transactions increases twentyfold. If detailed receiving and inspection entries are followed by three-way matching when the invoice is entered, the company will have to either hire a lot of people or find better ways to approach the issue.

The traditional approach is based on having the buyers as the only liaison with the vendors, having to double-check vendors to ensure they are not cheating, and receiving large batches with no regard to timeliness. Fortunately, as the procurement process is simplified, the systems associated with it can also be simplified. The ideal is to have a broad contract with the vendor and then call off each day only the materials and components required to complete today's production. There would be no receiving process and no inspection because the relationship is such that vendors can be trusted to deliver the right things at the right time with no rejects. Without a receiving process, there can be no accompanying complicated accounts-payable process. Ideally, the vendor should not have to send invoices because the production-completions and backflushing programs will identify the use of materials or components provided by each vendor and will create a payment record in the accounts-payable system. Indeed the payment could be made on a JIT basis using electronic funds transfer. Few organizations have truly achieved this level of waste elimination but this is the direction toward which they are moving.

Contract Purchase Orders

A contract purchase order negates having a buyer or purchasing agent be involved every time material needs to be ordered. The buyer will negotiate the terms of the contract on an annual (or longer) basis with the vendor. The contract will lay out the price, delivery conditions, and all other contractual issues but will not give specific quantities for delivery on specific days.

When material is required for production, the shop-floor personnel will contact the vendor and request delivery of the amounts required. This "calling-off" process can be achieved in many different ways — a call-off report, a fax, a *kanban*, or a simple telephone call. However, the key is that material is pulled from the vendor according to immediate production needs and not according to some buying schedule or economic order quantity.

The computer system must support blanket or contract purchase orders and must also be able to create a purchase order that does not specify delivery dates and quantities. The more flexibility built into the contract orders the better. It must be possible to specify the particular products for purchase and the total quantities the company plans to buy from the vendor over the life of the contract. Many vendors require this kind of commitment. It also must be possible to not specify particular products on the contract but just to lay out the agreement between the two companies so that production personnel can call off what they require when they require it. As the relationship between the companies becomes more and more cooperative, the contracts can become less specific.

Call-Off Reports

A call-off report informs the supplier of the amounts required to be called-off against a blanket order. In most cases, the call-off report will also provide forecasts for each item. The forecasts, although not firm orders, can be used by the vendor to plan future production and material acquisition. The call-off report can be initiated manually by production operators requesting a quantity of material on a specific day. The quantity and date constitute a firm order.

The call-off report can also be initiated by MRP. The MRP regeneration, when run, will create computer-planned orders for all components and raw materials showing quantities and dates (and possibly times) for delivery. These orders can be reviewed by the materials planners, and call-off quantities can be determined. The computer system must have a method of automatically converting computer-planned requirements to call-offs or releases against the contract purchase order. Some systems have a batch run that converts all materials required by an inputted date into releases and then prints call-off reports. Another approach is to have the conversion from recommendations to releases be a part of the review process. Again, the planners enter a "through date" — that is, the date through to which the orders will be firmed up — and the screen displays the items recommended by MRP. The planner can review and, if necessary, change quantities and dates. The system then creates the call-off.

Printed call-off reports will show not only the firm quantities but also forecasts of future requirements. These forecasts are taken from the computer-planned orders in the period beyond the through date used when firming the purchase-order quantities. In some industries, particularly the automotive business, call-off reports commonly have raw and fab dates associated with them. The "raw date" is the date out to which the customer will pay for the raw material the vendor requires to meet the needs of forecast production quantities. The "fab date," which is closer than the raw date, is the date through which the customer will pay for the fabrication of the forecast items even if they do not call them off as a requirement. This arrangement, common in some industries, gives the vendor that is not a JIT manufacturer a degree of financial security.

These call-off reports need not contain any firm orders. Many companies moving into agile manufacturing call off material from vendors manually and do not enter firm orders into the computer at all. A call-off report often is not practical or flexible enough for daily ordering of materials and is used primarily by companies ordering weekly or biweekly as a stepping stone to JIT. The call-off report can still be used when all the calling-off is done through manual systems because it can be used to convey the forecast information to the vendor. If the vendor is to provide JIT deliveries on a daily basis directly to the shop floor with zero defects, it is fair that they should be provided with good information concerning future requirements. This information can be conveyed through the use of the call-off report or a similar report with another name.

The call-off report does not need to be printed on paper. Increasingly, companies are using electronic data interchange (EDI) to transmit this information. Apart from the obvious increase in speed associated with the use of EDI, there are also enormous advantages in terms of waste elimination. EDI purchase orders can be automatically entered into the vendor's sales-order system without the need for someone to reenter information manually. This is a major time- and cost-savings factor. In addition, the lack of manual intervention means fewer errors and therefore greater accuracy. However, using EDI to automate unnecessary transactions is no savings. EDI should be used only to increase the speed and accuracy of the process.

Supplier Kanbans

As a company progresses with agile manufacturing, the calling-off of materials becomes a daily, twice-daily, or as-required procedure for the production operators. It is not necessary or desirable for them to have to enter the call-off into the computer system. A manual approach is usually more successful and often includes the use of supplier kanbans.

A supplier kanban is similar to a production kanban except that it is associated with an outside supplier rather than an internal production cell. When raw material or components are required, production personnel will send the kanban to the vendor who then delivers a kanban quantity. There is no distinction made between the operation of a supplier kanban and that of a production kanban. The fact that the material comes from an outside supplier makes no difference. The kanban itself will have different information printed on it to identify the supplier and delivery location, but usually the supplier will already have a standard routine for daily deliveries to the company.

A kanban does not need to be a physical card. Often the containers used to store the items can trigger delivery of the material. A fax can be sent or a telephone call made to the vendor. An electronic kanban can also be useful. Instead of sending a kanban card, production enters the required part number into the computer system and the requirement is made available to the vendor. The vendor can be notified immediately if there is a real-time network link between the customer and the vendor, or through an EDI link whereby the vendor will poll the EDI network on a regular basis to pick up orders from customers.

Computer systems used in agile manufacturing must be able to print supplier kanbans, based on the current production schedule, that can be used to call off materials from the vendor. EDI and electronic kanbans should also be supported by the system when there is a need. A particular advantage to computer systems using the electronic kanban approach is that call-off transactions are recorded in the system without the additional effort of having to be entered. If manual call-off or kanbans are used, the call-off information must be derived from elsewhere. It is counterproductive to require entry of kanban call-off information into the computer system if it is not required for another purpose.

Supplier Certification

Some larger organizations with many suppliers have additional computer systems to support the supplier certification process. These systems are very specific to the needs of the individual organization. In most cases, nothing specific is required to support the certification process because there are few enough suppliers to make a manual administrative system the most effective.

However, being able to track components and raw materials that are subject to certification is often useful. This information is not only good reference information for decision making but can also be used to determine how the receiving process should be conducted. If a receipt recorded for an item also has a certification code showing that no inspection is required, the system will recognize this code and post the item directly to raw-material inventory.

World-Class Purchasing

Changes in procurement are not the starting point of JIT manufacturing. However, as a company reduces cycle times and batch sizes, the need for JIT deliveries becomes apparent. In addition, eliminating purchasing, receiving, and accounts-payable processes is an important element of waste reduction.

Step 1. Use contract or blanket orders with call-off reports.
Step 2. Introduce call-off from shop-floor cells.
Step 3. Eliminate inspection and dock-to-stock activities through standardization and certification.
Step 4. Eliminate three-way matching.
Step 5. Eliminate invoicing by creating automatic invoices through the receipt transaction.
Step 6. Eliminate the receipt and invoice process by creating transactions through backflushing or EDI ASNs.
Step 7. Pay vendors just-in-time through electronic funds transfer.

Remember that the entire procurement process is non-value-added and should be eliminated. Software must have the flexibility to provide detailed tracking in the early stages and then allow for the elimination of most procurement activities.

Certification information is held at supplier and product level. Most procurement systems have a supplier-product table for vendor-specific prices, vendor descriptions, and part numbers. This same file can be used to contain the certification information. The certification information will show at least the level of certification and the certification reference associated with the documentation raised as a part of the certification process.

When a company is going through the certification process with a number of suppliers, the computer systems must keep track of the information required for assessing certification, including delivery and inspection details. Although the objective is to eliminate these processes, there is still a need for them to be available for use with vendors in the process of being certified. Even after certification is complete, auditing vendors to ensure that problems are not developing is sometimes necessary. Usually these audits can be controlled manually and do not need to be included in the computer systems. However, if the functionality is already available and can be used without confusion, then it is helpful to use it.

Eliminating Receiving Transactions

Using blanket or contract orders combined with call-off from the shop floor eliminates wasteful procurement and purchase-order processes for production components and materials. The next area of waste is the receiving and inspection process itself. The inspection process can be eliminated by certifying suppliers and by directing delivery to the shop floor. If the materials are provided by certified suppliers, there is no need to inspect the items and no need for complex inspection systems to accompany the receiving process. Eliminating incoming inspection eliminates the need for software support of that activity.

During the certification process itself and during audits of certified vendors, providing support for the inspection process will be necessary, even if this inspection consists merely of reviewing the statistical-process-control (SPC) charts and other certificates of conformance information provided by the vendor. The software does not need to be as complex as that of a traditional manufacturer because the agile company does not have the elaborate dock-to-stock procedures. In addition, delivery cycle times are so much shorter and quantities so much smaller for an agile manufacturer that errors or problems can be identified and resolved immediately. The complex tracking systems used by traditional manufacturers for the incoming inspection process are eliminated.

Elimination of the receiving process is more difficult to achieve because of associated financial transactions. Most purchase orders stipulate that the product becomes the property of the customer when the material is received. Most companies have problems with their auditors if they do not complete a receiving transaction showing the receipt of materials from their suppliers. However, these transactions can be eliminated if production cycle times are short. If the cycle time is short, if the vendors deliver on a JIT basis, and if the company employs single sourcing of materials, then material receipts can be included as part of the backflushing process. If the finished product has been made, then the material must have been used. If the material has been used, the company must have received it. Because the material is single-sourced, it must have been received from the certified vendor. Therefore, a receipt transaction can be created automatically showing the material as received on that day.

Clearly, this kind of approach can only be used when there are short cycle times because the time of receipt and the time of product completion must be closely matched. Similarly, there must be no work-in-process or raw-material inventory. There must be a clear one-to-one match between what was used and what was received.

An electronic kanban is also useful in this area because it can act as a receiving transaction. Provided the delivery lead time offered by the vendor is short and the vendor is certified, always delivers exactly what is ordered, and delivers it on time, the electronic kanban quantity can be used as a receiving quantity. When EDI is used and the vendor transmits an order acknowledgment or an advanced shipping notice, and when supplier lead time is short, then the EDI transactions can act as a receipt transaction.

Because it is important to eliminate as many transactions as possible, software support systems must have the flexibility to use these other kinds of transactions to create receipt transactions. Of course, the traditional receiving process may still be needed. It will be used for items purchased irregularly from noncertified suppliers, for occasionally auditing certified suppliers, and for new products and projects not yet a part of the certified vendor program.

The traditional receiving process used within agile manufacturing needs to be simplified. Using blanket or contract purchase orders and single sources of supply eliminates the need for receiving people to identify a receipt to a specific purchase-order line or date. There will be only one purchase order for the item and only one supplier. Therefore, the system will be able to process the receipt transaction from a minimum of information; it needs only the part number, quantity, and date. The computer screens and forms used in the receiving process can be simplified because there is no variation on the receiving process. This simplification reduces the time taken to receive the product and eliminates errors. Errors can be further eliminated by bar-coding the shipping documentation or the material's packaging.

Tracking by Lot Numbers and Serial Numbers

Tracking by lot number and serial number has become widespread in recent years as legislation and regulation have expanded into a wider range of industries. While only pharmaceuticals and foods formerly needed detailed lot- and serial-number traceability, now almost every industry is

affected. Lot- and serial-number control requires more transactions as well as more detailed tracking of inventory movements. While counter to agile objectives, the requirement is
unavoidable in many industries.

Full-lot (or serial-number) traceability requires the
manufacturer to identify the individual lots of raw materials
or components entering the production plant. As material is
issued to the shop floor for production, it must be possible to
trace which lots of raw material were used in which lots of
finished product; and then which lots of finished products
lots (or serial numbers) were sold to which customers. This
way defective components or material can be traced back to
every customer.

As discussed in Chapter Four, systems that track lots
and serial numbers have a structure similar to a bill of materials while keeping a separate bill for each lot or serial number.
This detailed information must be entered into the system
throughout the production process. The receiving process will
include the requirement to enter lot numbers or serial numbers for materials and components that are designated lot or
serial controlled. If the system has been simplified to eliminate the detailed receiving transactions, then the shop-floor-
completions reporting program will be the point at which lot-
or serial-number information is entered. The completions
program, when using lot traceability functions, requires the
entry of lot or serial numbers for items flagged for lot or serial
tracking.

Recording lot or serial numbers at the receiving stage
is not necessary providing the information is readily available
at the time the detailed information needs to be entered into
the system. Suppliers may have to use printed labels or bar
codes to identify the lot or serial number. Agile manufacturing now recognizes traceability as a requirement and devises

the simplest method for tracking this detailed information when required by customers or regulators.

Eliminating Three-Way Matching

Traditional companies use three-way matching to check that the supplier is not overcharging and to catch transaction errors in the company or vendor processes. The three ways are the purchase order, receipt, and invoice quantities. An agile manufacturer eliminates any need for these checks and balances by simplifying and perfecting the procurement and receipt processes and by creating close working relationships with vendors. Three-way matching is wasteful. Accounts-payable personnel are required to enter invoice information in detail — information that was previously entered when the receiving transaction was made. The resolution of discrepancies is a time-consuming and often finger-pointing exercise that is entirely non-value-added and does nothing to build the team cooperation so important within an agile approach.

The first step in eliminating these processes is simply to switch off the three-way matching procedures. Most integrated purchasing and accounts-payable systems have a three-way matching process. Most of them also have a parameter to allow the three-way match to be switched off. This eliminates the necessity for accounts-payable personnel to enter the detailed invoice information because there are no matching discrepancies to research and resolve.

Once this switch-off has occurred, the accounts-payable processes can be simplified further by the use of automated invoicing. Invoicing can be automated by having the backflushing process create invoices for items used in the production of finished products; by accepting an EDI transaction from the vendor stating what has been dispatched and thus automatically creating an accounts-payable invoice entry from this transaction; or (if a receiving transaction is required for another purpose) by having the invoice created from the receiving transaction instead of being matched against it.

The skeptical accountant or auditor will state that the purpose of three-way matching is to uncover errors and discrepancies in the process and that its elimination will allow inaccurate accounting records to go unchecked. This accusation would be true in a traditional manufacturing environment. However, an agile manufacturer simplifies the process and introduces fail-safe practices that virtually eliminate the possibility of serious error. In addition, the vendor-customer relationships developed as part of agile manufacturing render checks and balances unnecessary.

However, the most important aspect is that materials received each day are used the same day, allowing any errors to be detected, identified, and rectified that day. Pricing of materials does not change because prices are set by contract orders; shipments occur every day, so freight costs are standard or consistent. The sheer repetition of events and the simplicity of the process eliminate the need for checks and balances. In addition, the fact that suppliers provide zero defects means that there is no need to track and validate returns against invoice amounts. If they do occur occasionally, they can be handled manually. All these improvements and efficiencies add together to perfect the process of purchasing, receipt, and invoicing.

Short Cycle Times, Zero Defects, and On-Time Deliveries

Note that these approaches are possible only when there are short production lead times, zero defects, and on-time deliveries. The various aspects of agile manufacturing all work together to make the entire process of short cycle time and 100 percent-quality JIT production a reality. Abandoning important tracking of company assets (like inventory) and liabilities (like accounts-payable invoices) is irresponsible if the shop-floor changes have not been made and vendor relationships have not been created.

Some people call agile manufacturing a whole or holistic process. Each activity and change must be supported by symbiotic activities and changes in other aspects of the company's business. JIT manufacturing requires short cycle times and on-time deliveries with zero defects. Eliminating wasteful receiving and accounts-payables tasks requires short cycle times and on-time deliveries. Agile manufacturing requires eliminating waste in every aspect of the company's business in order to create excellence. Nothing can be ignored or neglected.

Summary

Radical changes are required in vendor relationships and procurement activities as a company moves into agile manufacturing. Purchasing's primary objective in a JIT environment is to nurture suppliers who can provide on-time deliveries and zero defects. This goal cannot always be achieved immediately and requires close, mutually beneficial, long-term relationships with a reduced number of vendors. While the use of single sourcing and vendor certification are mechanisms for creating a world-class approach to procurement, success depends upon the relationships established with vendors.

Software support must substantially simplify the traditional approach to procurement, which is largely based on old ideas of adversarial company-vendor relationships. Blanket or contract purchase orders eliminate the need for buyers or purchasing agents to place purchase orders. Calling off material from the shop floor must be simple, fast, and straightforward. The use of supplier kanbans, faxed orders, or electronic signals simplifies the call-off process. Software must support a variety of procurement approaches as vendor relationships develop and processes become simplified.

Another important area of simplification is the complex process of matching receipts and invoices. These checks and balances must be eliminated as world-class relationships are created with vendors. As always, the objective is to eliminate wasteful and unnecessary transactions — including receipts, inspection, dock-to-stock tracking, invoice matching, and invoicing itself. Invoicing can be replaced by the automatic creation of invoice records from backflushing, advanced shipping notices (ASNs), or receiving transactions.

CHAPTER EIGHT

Performance Measurement

T he need for changing production processes when moving into agile manufacturing is well understood. While there is general agreement about the importance and effectiveness of changes that fall under the categories of total quality management (TQM) and just-in-time (JIT) manufacturing, there is neither agreement nor consensus about the need to change a company's performance-measurement system and management-accounting methods. This chapter explores both the reasons for introducing a new approach to performance measurement and the characteristics of a new measurement system.[1]

The New Performance Measurement Systems

As companies move to agile methods, new performance measures are needed to control production plants. Old approaches to performance measurement do not work in the new environment. There are three primary reasons why new performance measures are required:

1. Traditional management accounting is no longer relevant or useful to a company moving toward an agile manufacturing environment.
2. Customers are requiring higher standards of quality, performance, and flexibility.
3. Management techniques used in production plants are changing significantly.

Traditional methods of management accounting and their accompanying performance-measurement systems were developed during the late nineteenth and early twentieth centuries. By 1930 all of the management-accounting techniques had been developed and standardized. Although there have been dramatic changes in manufacturing techniques and technologies over the past 60 years, management accounting has stayed the same. A gap has developed where many management-accounting techniques and concepts run counter to the needs of agile manufacturers. This conflict leads to company managers being misled because the accounting system is measuring wrong things in the wrong way. People are motivated to do the wrong things because they endeavor to achieve irrelevant targets. These problems with traditional management accounting are discussed in more detail in Chapter Nine.

If manufacturers were asked if they find customers more demanding now than five years ago, they would all answer yes. It was once acceptable to be less than prompt with deliveries — but this is no longer the case. Rejects were also once acceptable — indeed, most companies included reject rates in their material planning systems to accommodate for suppliers' lack of quality. This also is no longer the case. Customers have become more demanding in recent years for several reasons. One is that as many customers themselves move into world-class manufacturing, they require more reliability. Another is the consumer trend for more choice and better service. This trend gives a significant competitive edge in some industries.

The trend toward single-sourcing has also affected customer expectations. With more and larger orders going to fewer suppliers, the customer can make choices based upon quality, reliability, and long-term relationships. New methods of measuring production, quality, and distribution performance are required to monitor and meet the stringent demands being made upon an agile manufacturer.

Management issues stem from a fundamental change in philosophy toward people within the organization. An agile company creates cooperative working teams, fosters employee involvement, and moves authority and responsibility to the shop floor, office, and warehouse. These changes are intended to harness the skills and creativity of the entire workforce toward continuous improvement, high quality, short cycle times, and improved customer service. In this environment the reports traditionally used by middle managers are no longer useful. New methods of reporting performance are needed — methods that are relevant and timely for the shop-floor operators.

Characteristics of the New Performance Measures

In recent years many companies have experimented with performance measures that augment and encourage their world-class approach. However, while most of these performance measures have been in use for many years, what is new is the importance attached to them. These measures truly drive production and distribution processes, replacing traditional accounting and variance reporting.

Although performance measures used by agile manufacturers vary considerably, the following seven characteristics are common and should be taken into account when developing a new set of measures:

1. The new measures must be directly related to manufacturing strategy.
2. They primarily use nonfinancial measures.
3. They vary between locations.
4. They change over time.
5. They are simple and easy to use.
6. They provide fast feedback of information.
7. They foster improvement (not just monitor it).

Directly Related to Manufacturing Strategy

Proponents of agile manufacturing tend to clearly define their manufacturing strategies. These strategies are thought-through statements of the role manufacturing plays in the company's approach to its customers and its market. Business strategies vary significantly from one company to the next, even within the same market, according to the kind of company leaders want to develop. The manufacturing strategy of a company seeking to be on the leading edge of technology will be quite different from that of a company addressing the same market through low prices and generic products.

An agile strategy will focus on such issues as quality, reliability, short cycle times, flexibility, innovation, customer service, and environmental responsibility. Manufacturing strategy will also be congruent with business strategy. The former is an element of the latter, and performance measures must be in line with manufacturing strategy.

There are two reasons for keeping performance measures in line with manufacturing strategy. The first is obvious — a company needs to know how well it is achieving goals laid down by the strategy. It is important to choose a small number of pertinent measures that enable company managers to assess their progress. A second reason is that people concentrate on whatever is measured. A firm that measures and reports the results of someone's work motivates that person to improve. In a real sense, the choice of performance measures can steer company direction. Appropriately selected performance measures give a clear signal to everyone in the company what the company's priorities are.

Nonfinancial Measures

For traditional manufacturers, financial results are of paramount importance in measuring company performance. Within agile manufacturing, other performance measures are of at least equal importance and are the only measures used by the operational staff. A distinction must be made between financial accounting and management accounting. Financial accounting is concerned with presenting a fair representation of the company's financial situation to outside parties. These parties include stockholders, the Securities and Exchange Commission, and the revenue services. Financial accounts are for external reporting. On the other hand, management accounting is used for internal reporting and is not governed by rules and laws connected to financial accounting. Management accounting is the servant of the operational managers and should provide valuable and useful information for the day-to-day running of production facilities. In reality, many companies regard management accounts as an extension of financial accounts and provide no useful service to operation personnel.

The fundamental flaw in using management-accounting reports for operational performance measurement is the assumption that financial reports are valid and relevant in the control of daily business operations. This assumption is wrong. Financial reports are not only irrelevant to daily operations, but are generally confusing, misleading, and often harmful.

To be relevant, performance measures must be expressed in terms that directly relate to the issues contained within the manufacturing strategy. Disguising the results in monetary figures is confusing and unhelpful. Issues must be measured directly in ways that make sense to people on the shop floor who need the reports to make improvements. If measures are expressed in financial figures, the first thing the operations staff must do is "translate" them back into something relevant and understandable. If you are interested in improving quality, measure quality — not cost of quality. If you are interested in measuring cycle time, measure cycle time — not labor and material variances.

Variation Among Locations

A notable aspect of agile manufacturing is that implementation of these radical changes varies considerably from one location to another. Products manufactured in different plants may differ. Different locations may service different kinds of customers. For example, one plant may manufacture predominantly military equipment because the primary customer is the government. Another plant may manufacture consumer goods for retail. Because products and customers differ from one plant to another, performance measures likewise must differ.

There may be cultural differences between the company's locations. A plant in Maine may be quite different from a plant in California or Mississippi. The people are different; their work habits and education levels may be different. If these differences are true within the United States, then they are amplified when a company has plants in different countries. For example, cultural and economic differences between a plant in Switzerland and a plant in Pakistan are significant even if they manufacture the same products. To attempt to have a single standard set of measures for these plants makes no sense. To devise a standard set of measures is possible, of course, but it would be misleading and irrelevant because the standard would not take account of significant differences between locations.

Traditional performance-measurement systems tend to be consistent across all locations. In fact, many companies pride themselves on having a single measurement system that enables realistic comparisons of the efficiency of one plant against another. Such measurements are not only misleading, but the comparisons are harmful. An agile manufacturer builds teams, and teams do not flourish when judged against each other. Agile manufacturers are concerned with trends rather than the actual values being measured. The concept of self-directed work teams and continuous improvement is that, through the joint efforts of the entire workforce, all operations aspects will improve over time. The trend toward improvement is the key — not how one location stacks up against another. The way to achieve improvement and build teamwork across locations is to encourage a free flow of ideas, processes, improvements, and people between all locations. This will not occur when there is judgmental rivalry between locations.

Another important difference when a company has more than one location implementing agile manufacturing is the role of the "champion." Experience shows that radical change always requires a champion. A champion is the person in the organization who takes charge of the change process and drives the improvements through the plant with vision, determination, and skill. If a company has two plants and two champions, there is little chance that the champions will see eye-to-eye on the issues. Both will approach the implementation of methods differently. Both will be convinced that their approach is right and the other is wrong. There is often animosity between the champions. While their commitment makes them successful, it likewise makes them resistant to conforming to another person's approach.

If a company attempts to clamp a standard performance-measurement system onto a plant in the process of transition to agile manufacturing, one of three things will occur: Either the champion (1) will leave the company in frustration, or (2) will acquiesce to the company's desires and his or her enthusiasm will be squashed, or (3) will find a clandestine way around the company's standard system. This situation results in lip service to the standard system while the champion pursues personal ambitions. None of these three scenarios achieves the company's objectives. Management must acknowledge the significant differences between locations and must provide different measurement methods designed to meet each location's needs.

Changes Over Time

Continuous improvement is a cornerstone of agile manufacturing. Radical change occurs when a company embarks on implementation of agile methods by overturning old approaches and introducing new ones. Although these initial changes and improvements are significant, important, and beneficial, it is the long-term changes that really move an organization into world-class status. Included in this longer term are the ideas of employee involvement. Employee involvement brings the entire workforce in line with world-class goals, gives real authority and responsibility to operators and clerical staff making the products and maintaining the process, and creates a structure in which long-term continuous improvement can flourish. Continuous improvement is not just a catch phrase. It is a way of life within agile organizations and requires significant employee involvement.

The starting point of continuous improvement is the quality circle, improvement team, or action group. (It does not matter what they are called.) They are cross-functional teams set up within the organization for the purpose of improvement. The teams have the authority to make changes;

they are self-directed and are responsible for continuously improving the quality, process, and products within their area of responsibility. They may be set up within a specific working area or to address a general issue. Successful companies find that flexibility is required in the way teams are established and in the missions they receive. The end result of team activities is to bring about improvement, and this improvement generally occurs in small steps. One objective of quality circles and employee involvement is for continuous improvement to become a way of life within the production plant.

Continuous improvement means the company deliberately has set out to instigate long-term institutionalized change. In all likelihood, as this change occurs, performance measures will need to change also. When world-class methods are first introduced, measures will emphasize the most significant areas of change. Other areas will become significant as implementation progresses and the emphasis will change accordingly. Similarly, customer needs may change over time. At one moment customers may demand on-time deliveries, making delivery reliability the focus of attention. Later, customers attention may move toward innovation or quality. Measures will still be needed to support on-time delivery objectives, but emphasis can now shift as customers introduce new performance criteria. Performance-measurement systems must have enough flexibility to adapt to the reporting emphasis as changes occur over time within the company.

Simplicity and Ease of Use

The most effective performance measures are ones people readily understand. Past tendency has been to devise complex measures that relate more than one aspect of performance into a single ratio or index. These complex or composite measures tend to be unsuccessful within agile manufacturing. To be successfully motivated by performance reporting, people must clearly understand the reports and be able to see the relevance to their jobs and the company's manufacturing objectives.

Plain and simple measures of the business's most important elements are better than complex and subtle measures. If an issue is measured directly and presented in straightforward terms, people find the results easier to use and the performance measure is more effective. Agile performance measurements are clear and direct. Traditional measurement systems rely primarily on reports distributed to managers who need the information. These reports are produced weekly or monthly and contain complete analyses of the issues being monitored. In contrast, results of performance measurement can be shown by more immediate and direct methods. These methods include the use of charts, graphs, signals, and bulletin boards.

Agile plants often display the results of performance measures continuously throughout the day on boards, charts, or graphs located adjacent to production cells or lines. The advantage of this kind of presentation is that information is being shown clearly, directly, and in a way that everyone understands. Direct reporting methods can be useful motivators because shop-floor personnel are able to monitor their own performance continuously and find the results displayed clearly for all to see.

There are many ways that this kind of direct and simple performance measurement can be achieved. Manually posted charts and boards are common; automatic monitoring through process-control systems is helpful when those systems are being used. Simple spreadsheet programs and graphic displays are opening up new approaches as the cost of personal computers (PCs) drops. The use of exception reporting through the computer systems and automatic E-mail messages can also be helpful. The media is less important than the message. To be effective, performance measures must be clear, direct, and simple.

Fast Feedback of Information

In most traditional companies, cost-accounting reports are available weekly or monthly and show variances for such items as material cost, material usage, labor productivity, labor rates, and overhead allocation. By the time these reports reach anyone, little can be done about problems they have exposed. Either the problem occurred so long ago that it is impossible to investigate the cause, or the problem has already been identified and corrected by other means.

An agile manufacturer must detect and resolve problems as they occur — not several days later when the reports are produced. Many problems can be detected on the spot by operators with the training and equipment that enable them to continuously monitor quality, flow rate, setup times, and other aspects.

A primary purpose of performance measurement is to assist people with problem solving, waste elimination, and continuous improvement. Timely information is essential if these objectives are to be met. For such issues as quality, production rate, or scrap, information needs are continuous. For things like customer service or schedule adherence, daily or twice-daily reporting may be appropriate. The timing of the measure must be appropriate to the issue under examination. For example, issues like time-to-market need to be reported over a longer time frame.

Fast feedback can be provided by using continuous measurement, either manually through charts, boards, or signals, or automatically through a data-capture system. Shop-floor terminals enable people to run reports and inquiries as needed. Having fourth-generation programming tools (4GLs) at hand for operations people opens up information that is available within the computer systems.

Fostering Improvement — Not Just Monitoring It

This last characteristic of agile performance measurement deals with motivation. Performance measures need to show clearly where improvement has been made and where more improvement is possible, rather than to merely monitor people's work. Traditional performance measures are based upon the concept of monitoring work so that people can be assessed, rather than of providing information to help people improve. This subtle but important distinction underlines the change in approach to managing people in a world-class company.

Many of the issues discussed previously contribute to this aspect of a performance-measurement system. Ease-of-use, fast feedback, and nonfinancial measures all make for measures that help the improvement process. Visual charts and graphs are better than computer-generated reports. Having people manually create their own measures enables them to take a more analytical view of the data. Measuring issues that relate directly to the company's manufacturing strategy ensures that people and their improvement efforts are focused upon achievement of strategically important goals.

To some extent, this aspect of a performance-measurement system relates more to how the measures are used than how they are designed. Even in traditional companies the measures themselves are not intended to belittle people — how the measures are used makes the difference. Traditionally, this has been to monitor individual work performance, often with a view to assessing pay and promotion. This practice leads to fear and nervousness rather than to innovation and improvement. Performance measures in an agile organization must be nonthreatening. Their purpose is to initiate improvement, build teamwork and cooperation, and allow everyone to become part of the decision-making and innovation process as the company moves into world-class activities.

Performance Measurement and Computer Systems

Traditional performance measures have been predominantly computer-produced reports. Agile manufacturers make more use of manual performance measures than traditional companies, but they also make use of information derived from complex systems like automated process control. There is a subtle balance to be struck because the use of computer systems for agile performance measures can be very sophisticated. Likewise, the role of the management-information-systems (MIS) department is crucial to the success of a new measurement system even as the use of computers for performance measurement changes significantly.

Producing Performance-Measurement Reports

For agile companies, performance measurement emphasizes graphic presentation and employee involvement. Hand-drawn charts at each work cell are common. Ideas of the "visual factory" are often employed, using boards and signals to indicate such things as quality, on-time delivery, and production schedules. Considerable benefit is gained by having those who use the information gather and present the information, and widespread use of statistical-process-control (SPC) charts is one element of this approach. These methods contribute to team building and the transfer of responsibility from middle managers to production supervisors, operators, and self-directed work teams.

There are limitations to manually produced performance measures. Information that is produced manually is difficult to distribute widely. Collating and summarizing such information, as well as any additional analysis, must also be done manually. Gathering data to create the chart must be quick and easy to do. There is no benefit if operators have to spend a considerable amount of time gathering information that is already available on a computer system.

In general, computer systems should not be introduced solely to gather data for performance measurement. Its use does make sense, however, when information is being gathered for another purpose. In many companies the computer systems gather process-control information, production completions, inventory transactions, quality data, and so on. Being gathered already, such information can also be useful for performance measurement.

Shop-Floor Computers and Terminals

Traditional companies tend to separate computers and terminals from the shop floor. They put them in the production-control office or — worse yet — in the data-entry department. In many companies, information is still entered into the computer by data-entry clerks instead of by the people actually responsible for its accuracy. An agile manufacturer would readily identify such practices as "waste" and eliminate them. The first step is to eliminate as many transactions as possible and then have people on the shop floor or in the office enter their own transactions. This system is far more accurate and puts responsibility in the right place.

With personal computers becoming more affordable, they are more commonly seen on the shop floor. These PCs initially are used either as terminals linked into the mainframe system, as part of the computer network, or as a server in a client/server configuration. Shop-floor PCs can be much more than just terminal emulators. Once people are trained in the use of computers and software like spreadsheets and graphics, PCs can become valuable performance-measurement tools.

Information entered for transmission to the host computer can often measure schedule completions, production rates, quality, completions, and other relevant factors. If entered into a simple spreadsheet program, this data can be analyzed to provide important information about the production cell. Trends can be established, some basic statistical information can be presented, and comparisons can be made against previous weeks or other products manufactured on the cell. These simple analysis tools can provide powerful and flexible information to people in the cell and to the improvement teams working on these issues.

If performance-measurement information is available within a shop-floor PC, then graphic results can be presented quickly and easily. All leading spreadsheet applications contain built-in graphics packages designed to quickly provide clear and effective graphic presentation of information. Thus, operations personnel can use spreadsheets not only for analysis but also for presenting information. Looks are irrelevant. A simple, hand-drawn chart, while less sophisticated, is often the best way to present information. The usefulness of simple PC spreadsheet programs is in the speed and accuracy with which they provide analysis and graphics.

Another asset of having PCs on the shop floor is their ability to consolidate information. One problem with manually produced information is that consolidating it for broader operation is difficult and time-consuming. This consolidation can be done more easily when the information is contained within a PC. If every work cell is collecting information on quality, production completions, production rates, and the like, then it is simple to bring information from every cell together into another computer for analysis and measurement. If the PCs are linked together through a network or client/server environment, information can be consolidated directly. If the PCs are not networked, then data files from each PC will have to be loaded onto another machine for consolidation purposes. This consolidation can be done through a custom program or through a spreadsheet-style analysis.

This approach can be employed when statistical process control is being used on many machines and where SPC information is being collected through a PC or other computer. The results of each machine's SPC can be consolidated with all other machines and processes to create consolidated statistical and performance-measurement reports.[2]

Automated Process-Control Systems

Companies that use an automated process-control system on the shop floor can extract information from the system for the purposes of performance measurement. This information may be presented continuously or summarized over a period of time. Some process-control systems are sophisticated, complex computer systems custom-designed to control and monitor an automated production process. Other systems are less complex and are used for monitoring and tracking one or two aspects of production.

If the process-control system is gathering useful performance-measurement data, then that data should be extracted and used. Introducing a system specifically to gather the information is wasteful. However, if the information is already available, using it is sensible. The kind of information often available is data relating to such things as quality, production rate, completions, and scrap and rejects. This information can be extracted and presented graphically or in the form of a chart or table. It can also be used to alert operators when problems occur that require attention.

An example of a simple but effective, automated process-control system is at Bloomington Seating Company in Normal, Illinois. This company makes car seats for Chrysler and Mitsubishi using a well-thought-out JIT cellular operation. When a set of seats is completed, the set is placed into a standardized container and routed by conveyer to an automated warehouse for consolidation and shipment. Containers are labeled with permanent bar-codes that are automatically read when the container passes into the warehouse area. This completion data signals the completion of the seat set and update the computer systems by creating inventory transactions and backflushing. They also keep track of how many seats are being made, of what the production sequence is, and of the rate of production flow. This information is readily available without additional manual data entry.

Fast Feedback to Operators

Providing fast and appropriate information feedback to personnel in the factory, warehouse, and office requires an opening of the data available within the computer systems. Many traditional companies do not allow access to data contained within computer systems. The computer and its data are the domain of the MIS department, which is often unwilling to allow access to other users. This approach changes as employee involvement progresses because improvement teams must have information if they are to be successful.

The first step is to create reports that are useful and valuable to the improvement teams. The second step is to give teams, operators, and supervisors the ability to create their own reports as needed. Most traditional companies make performance reports (and others) available when it is convenient for the MIS department. On the other hand, a company focusing on improvement and employee involvement will make the information available when people need it — they will be given the authority to run the reports at any time according to their needs.

Certain problems are associated with this approach. Some reports are only relevant after other activities have been completed. For example, a customer-service-level report is valid only after all of today's shipments have been entered into the system. To solve this problem, the people who need the report must be aware of the limitations and must check that all activities are complete before the report is run. In many cases, these same people will themselves be involved in completing the other activities.

Another problem is that having people run large reports at any time during the day can affect a computer system's efficiency and response time. Many companies run major reports overnight so that the computer is fully available for data entry during the day. Technical problems may arise with running a report while other people are accessing and updating data. A report may require access to a file within the system and may prevent other people from accessing that file at the same time. This safeguard is common practice within the data-security functions of many computer systems. These issues must be explained and people must be given the authority to ensure that such issues are understood and considered.

In reality, when authority for running reports is given to those who use them, the result is that fewer reports are run because they are run only as needed. However, the burden on the machine often increases because reports are run at peak times instead of overnight or at another off-peak time. Some companies compromise and schedule time during the day — for example, lunchtime — to make these reports available. Other companies recognize the importance of the issue and obtain a more powerful machine. As computers increase in power and as their prices fall off, it becomes feasible to make a large improvement in computer power with a relatively small investment.

Fourth-Generation Report Writers

An extension of the philosophy that information contained within the computer should be readily available to the people within the company is the use of fourth-generation (4GL) report writers. A 4GL is a computer program that allows the user to easily extract information from the computer data base and print customized reports. These reports may be used just once for a particular project or analysis, or they may be added permanently to the new performance-measurement system. The key to a successful 4GL is the ease with which people can become proficient at creating their own reports without having to be trained as computer programmers. The 4GL must handle internally any technical aspects of designing and programming a report so that creating the report is simple.

When a good 4GL is available within the production plant, information within the computer system can be accessed and used. One or two people within each area can be trained to use the 4GL. New required reports can be created using the 4GL instead of having the MIS department write a program. Most companies overload their MIS departments. When an operator or improvement team needs a new report, MIS personnel often have great difficulty providing the new program as quickly as needed. This situation often leads to conflict between MIS and operations personnel. A 4GL solves many problems because operations people have a tool that enables them to create the report themselves.

There are some significant problems connected with the use of 4GL report writers. One is that 4GLs tend to be "resource hogs" — a 4GL program often takes a considerable amount of computer resources. Despite being easy to use, a 4GL is a large and sophisticated program requiring a lot of memory and computing power to be fast and effective. A computer has only a certain amount of power and resources. Thus, if too many people are using the 4GL reports, the computer system can be slowed down significantly thereby hampering other users in their work.

Coupled with this resource issue is the problem of the proliferation of reports. If every area can produce its own reports, there is a tendency to run a huge number of reports daily or more often. Many will be very similar because the needs of different departments within the same plant will be similar. Large numbers of reports create confusion because the same information is being reported many times over from slightly different viewpoints. This overlap can also create conflict between people who look at the same information differently.

Therefore, some coordination of 4GL use is required. Its power in making information available throughout the organization also creates confusion and problems with resources. Coordination would involve reviewing the use of 4GL reports and rewriting the commonly used ones in a conventional programming language. Conventional languages like "C" or COBOL can create much more efficient reports than a 4GL report writer. Another coordination aspect is to review the reports to limit overlap. If similar reports have been written in several areas, it is better to create a single report that meets the needs of each area. While coordination is required to overcome the problems caused by 4GL report writers, care must be taken to ensure that the coordination itself does not become a wasteful bureaucracy.

Features of a Good Report Writer

To be valuable, a report writer must be easy to use and powerful. MIS people have difficulty selecting a report writer because the ease-of-use aspects must be viewed from the user's viewpoint. Generally, users will not have a background in MIS or understand file structures or programming procedures. A good 4GL report writer is designed to be a helpful tool for this kind of user. Some of the basic features of a good report writer include the following:

- *Simple instructions*
  Computer-programming languages are known for their confusing and complicated terminology. A 4GL should use plain English.

- *Prompted entry*
  Users should be prompted step-by-step through the process of developing reports and inquiries.

- *On-line error checking*
  Any mistakes should be identified and highlighted immediately so that the user can correct them. Some report writers do error checking at the end of the process, which requires less resources when using the 4GL but makes the program more difficult and complex to use.

- *"What you see is what you get" (WYSIWYG)*
  As the report builds up, the exact format should show on the screen allowing the user to see what the printed report will look like. This feature makes designing a usable report much easier.

- *Copy and modify features*
  Often a user will not write a new report from scratch but will modify an existing report. The 4GL should not require recreation of the entire report but merely modification of the existing one, or should copy it to a new name and then modify it.

- *Automatic formatting*
  In a traditional programming language, the most tedious and time-consuming part of writing a report is formatting — laying out which field goes where, where page breaks occur, and so on. This task must be done automatically by the 4GL.

- *Comprehensive data dictionary*
  The 4GL data dictionary contains and defines all files and data fields for use in producing reports. If the data dictionary is well structured and well presented, the 4GL is much easier to use. How the system files are set up and how the various files relate to one another is important when creating reports or extracting data.

- *Quick and simple ad-hoc reports*
  It must be possible to produce a simple report quickly. Even if the report writer has additional features and functions, it should be able to create a simple report without requiring the more elaborate process required for more advanced features.

- *Well-written user manual*
  Even though most people do not read manuals, having one that is both easy to use and also comprehensive is absolutely essential. The manual should give

specific examples, preferably with information about the specific system and files being accessed.

- *Thorough training package*
  People using the 4GL must be trained in its features. A good program can be difficult to use if people lack adequate training.

For the report writer to be an effective performance-measurement tool, the following advanced features must be available:

- *Multiple file access*
  Performance-measurement reports usually require information from more than one file within the manufacturing and distribution systems. Old report writers could only access a single file. Today's report writer must access up to at least ten files.

- *Several lines of data*
  The format of reports and inquiries must allow for more than one line to be printed from each set of records accessed.

- *Line and total-level calculations*
  The report writer must provide calculation functions at the line level, the subtotal level, and the total level.

- *Wide range of selection criteria*
  Data selection must be flexible and easy to use. The report writer should be able to select according to multiple-data elements and across ranges of values, and use full Boolean logic. It must be possible to enter the ranges "on the fly."

• *Wide range of mathematical functions*
The report writer should offer all mathematical and statistical functions required for selecting, analyzing, and presenting information.

• *Flexible totaling and subtotaling*
Totaling and summarizing are important aspects of performance-measurement reporting. The report writer must provide multilevel totaling with appropriate user-defined page breaks.

• *Offer various paper sizes and type faces*
Final output is as important as the report's information. The report writer should have considerable flexibility in the way final information is printed.

• *Suppress lines, headings, etc.*
While detailed information is often less important than the summarized totals, detail is required to calculate totals. The report writer must be able to access detailed information but suppress printing it on the report.

• *Have automatic graphic capabilities*
Visual presentation is a cornerstone of good performance-measurement systems. The report writer must be able to present information graphically on printed reports.

Other Fourth-Generation Features

Fourth-generation report writers often contain even more flexible and powerful features. The previous discussion considers only 4GL report writers. In fact, while some 4GLs are fully featured system-development tools that can be used to create large and complex computer applications, for the purpose of performance measurement, a less complex 4GL is needed. Other helpful features include:

- *Uploading and downloading data*
  This capability extracts data from the data base and downloads it to another machine or PC. This activity is useful when the 4GL itself cannot produce graphic presentations and data needs to be fed into a graphics package. The upload/download feature is also useful when ancillary analysis is required from a machine other than the primary machine containing the production planning and control system. This additional analysis may be related to such areas as engineering, quality, shop-floor layout, or human resources. Uploading data from a PC or workstation to the mainframe or central computers is required when results of some additional external analysis must be fed back into the primary systems. There are many reasons for feeding back information including sales forecasts, shop-floor scheduling, and engineering changes.

- *Subsystem development*
  Another useful feature is the ability to create small subsystems to support a performance-measurement system. Specific information required for a measure may not be readily available on the standard system, and a 4GL with subsystem development tools can be useful in providing this additional functionality.
  An example would be a distance-moved report that requires information about distances from one cell or warehouse to another within the plant. This information is not readily available from a standard production planning and control system. An additional file containing this information could be defined by

the 4GL and a report written showing the distance the material moved throughout the plant. To support this kind of small subsystem development, the 4GL must be able to define new files, integrate those files with standard files, update and maintain data in the new files, and produce reports from both the standard and the new files.

These kinds of tools are more powerful than a simple report writer, and their use demands adequate control. An inexperienced or careless user could do a lot of damage with tools that enable files to be updated directly without going through an application's standard data-entry programs. Most companies use these tools only within the MIS department in liaison with the users requiring the additional functionality.

Summary

Although performance measures vary considerably among agile manufacturers, they all incorporate seven common characteristics. These new performance measures:

1. directly relate to the manufacturing strategy
2. primarily use nonfinancial measures
3. vary among locations
4. change over time
5. are simple and easy to use
6. provide fast feedback to operators and managers
7. foster improvement instead of simply monitor

Although manual reporting of performance measures is common, the use of computer systems to provide performance reports still plays an important role. Data within the system is made available to operators and improvement teams for purposes of performance measurement. Shop-floor

terminals, local PCs or workstations for analysis or presentation of the information, and fourth-generation report writers
help achieve this goal. Fourth-generation report writers give
access to central computer-system information that can be
extracted to provide performance-measurement reports or
charts.

CHAPTER NINE

Accounting

F inancial-accounting methods employed by manufacturers are largely stipulated by regulatory agencies, government departments, and professional bodies. The way accounts look, posting methods, inventory valuation, depreciation, accruals, gross and net profits, and so forth are all laid out in time-honored accounting standards. These standards vary considerably from one country to the next, but typically include the Generally Accepted Accounting Practices (GAAP), the Federal Accounting Standard Board (FASB), the Statements of Standard Accounting Practice (SSAP), and the International Accounting Standards (IAS). Financial-accounting reports are for external reporting; they are required for stockholders, the revenue services, the Securities and Exchange Commission, and other government agencies.

These requirements do not apply to management accounting. Of course, there is no shortage of standards and directives, but management accounting is for internal reporting. Management accounts are the servants of managers and can be changed and eliminated according to organizational needs. Changing the methods of financial-accounting reporting is virtually impossible because they are required by external bodies. However, it is essential to change management-accounting systems when they no longer serve the needs of the organization.

Unfortunately, dismantling or radically changing any company's accounting systems is difficult. Some of the most advanced agile manufacturers continue to use traditional management-accounting systems despite their being misleading and harmful. There are three common reasons why management-accounting systems last beyond their usefulness.

One reason is the integration between the management accounts and the financial accounts. Like the parable of the wheat and the tares, you cannot remove management accounts without creating havoc with financial accounts because the two aspects are inextricably connected.

A second reason is that although a company may be achieving wonderful results with world-class and just-in-time (JIT) manufacturing, senior managers may still not have made the emotional transition from traditional thinking. They may still insist on seeing efficiency reports, variance reports, and budget-to-date spread sheets traditionally used to monitor the business.

The third reason is the intransigence of the management accountants themselves. Chapter Seven's discussion relating to the difficulties of persuading purchasing personnel to relinquish approaches they learned in school and have been using for years is even more true for management accountants. It is wrong to generalize because accountants in many successful agile manufacturers have been in the forefront of change. However, the majority of companies too often find that management accountants are rarely enthusiastic for change and often obstruct progress because they cannot see the need for change within their own departments.

In reality, the use of traditional management-accounting methods in a company moving toward agile methods is wasteful and frequently harmful. Many companies have failed to progress with the implementation of agile manufacturing and JIT production because the management-accounting reports showed negative signs. The problem is not with the methods — it is with the reports that measure the wrong things. Because traditional management accounting has been used for so long to gauge the success or failure of production operations, accountants and managers have difficulty recognizing that old practices are no longer relevant and can damage the business.

Problems with Traditional Management Accounting

Traditional approaches to management accounting recently have come under considerable criticism. When JIT manufacturing was first introduced in Western companies in the early 1980s, a number of writings highlighted problems caused by traditional management accounting.[1] However, the 1987 publication of the book *Relevance Lost: The Rise and Fall of Management Accounting* by Robert S. Kaplan and H. Thomas Johnson brought these issues to the forefront. Although the authors focused primarily on the historical development of management accounting and the problems of cost distortion through inappropriate overhead allocation, they also pointed out a number of other inadequacies of traditional cost- and management-accounting methods.

Over the last few years, as more and more companies have embarked on the journey to agile manufacturing, the inadequacies of traditional accounting methods have become generally understood. While there is no clear consensus on what should replace the outmoded techniques, everyone is agreed that change is required. Nonetheless, making these changes in a company is still a difficult task. Often computer systems are a part of the problem — being steeped in traditional thinking with regard to cost accounting. Their inflexibility makes it difficult to adapt to the organization's changing needs. This chapter examines how software can be harnessed to overcome traditional management-accounting problems.

Aspects of Traditional Management Accounting

The primary reasons for the existence of management accounting include the following:

- valuation of inventory
- tracking of work-in-process (WIP)
- identifying inefficiencies in production and support departments
- calculation of product costs
- management reporting and performance measurement
- preparation of budgets
- prevention of fraud

These requirements lead to the development of complex systems to provide the information needed to fulfill these management-accounting tasks. The systems usually involve:

- numerous cost and responsibility centers throughout the organization
- many transactions that primarily are entered manually
- complex and dubious allocation and assignment costs
- numerous variance reports

- a long and involved budget-development process
- time-consuming and expensive clerical procedures
- internal conflict resulting from the presentation of conflicting information

These complex and expensive systems could be justified if they provided valuable information. In reality, they provide information that is of questionable usefulness. In their *Harvard Business Review* paper, Thomas E. Vollman and Jeffrey Miller pointed out that every production plant contains two factories.[2] The first factory, which is visible, manufactures products. The second factory, which is hidden, processes transactions. Of the two, the hidden factory is far more expensive and complex and generates waste.

Why Are Complex Systems Needed?

Complex accounting systems are needed in a traditional manufacturing environment because inventories are high, cycle times are long, production processes are complex, and managers use financial reports to run the business. In an agile organization, inventory valuation ceases to be a major issue because inventory is low. The reduction of cycle times means there is much less inventory in work-in-process and fewer production jobs requiring control. The introduction of cellular manufacturing simplifies production to the point where there is no need for complex systems because the process does not warrant them.

The new manufacturing organization is characterized by cellular manufacturing, short travel distances, short cycle times, fewer departments and centers, simplified production processes, very high quality, very low inventories, a flexible and responsive team of production personnel, and nonfinancial performance measures. These changes eliminate any need for the traditional complex cost and management-accounting approach. The new approach to accounting systems limits the number of cost centers to the informational needs of production personnel; many costs are assigned directly, transactions are minimized, and the systems are simple, less costly, and more understandable to the people using them.

Relevance Lost

When techniques of management accounting were developed, the best method for automating calculations was the slide rule. Anyone using a slide rule today would be considered eccentric because changes in technology have rendered it obsolete. In the same way, changes in production technology have rendered management accounting irrelevant and wasteful.

To be relevant, cost and management accounting must support the company's manufacturing strategy. An agile manufacturer invariably has a manufacturing strategy that includes such aspects as quality, reliability, innovation, lead times, time-to-market, customer satisfaction, and social issues. None of these crucially important factors are monitored by traditional cost and management accounting methods.

Frequently, financial measures are not meaningful to the operational control of a production plant or distribution center. People in the factory do not need to think in financial terms; they concentrate on issues such as production rates, yields, on-time deliveries, rejects, schedule changes, and stock-outs. Cost accounting and variance reporting require that a production supervisor spend a considerable amount of time analyzing financial figures to make the information meaningful. These reports and processes are unhelpful and irrelevant.

Cost Distortion

During the early days of management accounting there was an emphasis on the detailed analysis of product cost elements. At that time, the cost elements were usually quite clear, labor by far being the largest (typically 75 to 95 percent) followed by materials and overheads. Cost patterns have changed markedly over the last 50 years. New technologies and production techniques have led to a reduction in the amount of labor required to manufacture a product. The labor content of an average product manufactured in the United States in 1988 was 7 percent of the total cost. In addition to the three primary costs, new elements are becoming increasingly important. These changing cost patterns mean that traditional cost-accounting methods concentrate on increasingly less important costs and ignore significant factors such as technology costs, engineering costs, and preventive maintenance.

Traditional methods for apportioning overheads are a major cause of cost distortion. The majority of U.S. companies use labor hours or costs as the primary factor for apportioning overheads. If labor content is low and overhead rates exceed 100 percent, 500 percent or more, total product cost will be significantly distorted. Some companies attempt to overcome this inconsistency by using other factors to apportion overheads; machine time or material cost, for example. While these approaches can be an improvement, the problem remains — when overheads are high and direct costs are low, distortion generates misleading cost information resulting in wrong and harmful management decisions.

Inflexibility

An important aspect of traditional management accounting is that the systems and approach are common to all parts of the company and do not change over time. This feature is seen as a virtue of standard accounting practices, but is harmful to the new manufacturing where flexibility is important. This flexibility manifests itself in a number of ways. Product volumes and product mix are flexible. Flexibility of methods and approach among plants differ within the same company. The importance given by WCM companies to continuous improvement requires that performance measures must change over time.

Traditional management-accounting reporting is usually governed by a strict accounting schedule based on needs of financial accounts and unrelated to needs of production plants. The schedule often makes the information available to production managers and supervisors too late to be useful. Much more flexibility and responsive operational control is required.

Integration with Financial Accounts

Do cost and management accounts need to be integrated with financial accounts? This controversy has created a raging debate among accounting professionals and academics and we are a long way from the last word on this subject. Financial accounts are required to present information about the company to outside parties. Management accounts are used internally by company managers to control, monitor, and guide the company's business. These two aspects of the accounts do not need to be integrated; they serve different functions and use different information.

Traditionally, the two aspects of accounting have been integrated with the cost accounts serving as a subsidiary ledger to the general ledger for purposes of inventory valuation, production-cost tracking, and so forth. An often useful agile feature is to have different methods of inventory valuation for internal reporting or no valuation reporting. The strictures of the financial calendar and monthly cycles hamper the usefulness of cost and management reports.

Impedes Progress

Traditional cost and management-accounting techniques run contrary to some key agile elements. There is an emphasis on the detailed collection of costing information, much of which is of limited usefulness. This activity is wasteful and should be eliminated.

Principles of cost and management accounting also entrench ideas about how to run the company that do not encourage the innovative thinking and continuous improvement so necessary for success in world markets.

Traditional methods used to assess the viability of capital projects (such as return-on-investment (ROI), discounted cash flow (DCF), internal rate of return, and payback periods) do not help when making an investment decision in a world-class plant. Agile companies are concerned to view the broader, long-term effect of these decisions rather than to emphasize one narrow aspect. Traditional techniques tend to point toward the purchase of large, high-volume, specialized equipment, whereas an agile manufacturer will be looking at such aspects as flexibility, fast changeover, small batches, and cross-functional training. Agile companies invariably make capital-equipment decisions based upon the usefulness of the new machinery or process to the implementation of world-class techniques like high quality, fast setups, and on-time deliveries. These are issues that matter.

New Approaches to Management Accounting

The previous examples illustrate substantial problems with the use of traditional management-accounting techniques in an agile work environment. How should managers and management accountants resolve these problems?

Simplification is the key. One approach is to eliminate cost and management accounting altogether. A second approach is to adapt some other tested methods to the new environment. These other methods need to be considerably simpler and more relevant than traditional methods. The third approach is to invent new methods that overcome the shortcomings of old-style management-accounting systems.

Eliminating Management Accounting

Some of the more radical companies have eliminated management accounting completely and control their businesses entirely with nonfinancial reporting. They do use some

---

Accounting Simplification: Phase 1

In most companies, simplifying the accounting systems can be achieved in a step-by-step process:

- Eliminate labor reporting.
  Stop detailed reporting of labor and machine times against individual production jobs.

- Eliminate variance reporting.
  This goal can be achieved by simply not running the reports.

- Reduce cost centers.
  Part of the complexity of traditional cost-accounting systems is the proliferation of work centers. As cellular manufacturing is introduced, this complexity is reduced.

financial reports, but these can be derived simply from the general ledger or from the completions and backflushing information within the production system.

Other companies maintain a management-accounting system but strip it down to a bare-bones minimum by eliminating the detailed reporting of labor hours, machine use, production job steps, and so forth. These changes can lead to the elimination of work orders, removal of variance reporting, and dismantling of integrated management accounts and financial accounts. Other tasks like detailed budgeting and month-end reporting can also be removed and replaced by nonfinancial, functionally based reporting and control mechanisms.

Computer systems that support the elimination of management accounting must be flexible. Usually, the elimination does not occur in one step; it is spread out over a series of steps that gradually reduces the complexity until the costing system is finally dismantled. If the cost-accounting system is not flexible and forces the user to adhere to a certain set of procedures, taking this step-by-step approach is difficult.

Process Costing

As a company implements agile manufacturing, production methods become more like those of a repetitive or process manufacturer. The company may still be a job shop making small quantities of different products, but the production process becomes more repetitive and more clearly understood. A great deal of time has been spent understanding and perfecting the process, and the implementation of cellular manufacturing introduces a "group technology" approach to production that fits well into a repetitive view of production. Repetitive or flow manufacturing lends itself more to process costing than to job costing.

Accounting Simplification: Phase 2

The second phase requires a little more depth of commitment and some software enhancements:

- Eliminate detailed job-step reporting.
  As cellular manufacturing is introduced and/or cycle times are reduced, it is no longer necessary to track by job step.

- Eliminate WIP inventory reporting.
  As cycle times are reduced and JIT manufacturing is created, work- in-process inventory decreases dramatically and the need for reporting it is no longer required.

- Employ backflushing.
  With reduced cycle times and cellular manufacturing, backflushing can be used effectively for reporting production completions and for performing component-inventory transactions.

Transition to process-style costing rather than job costing can be achieved smoothly because process costing is a standard textbook method of cost accounting. Implementing process costing does not require the radical changes needed to eliminate costing altogether, but does eliminate much wasteful data collection and analysis. Process costing is concerned primarily with trends in production costs rather than with the individual cost of a specific job. This practice is very much in line with a JIT and continuous-improvement approach.

Process costing collects costs by production cell and/or product or product group and presents the information on a month-to-date or year-to-date basis. The application of direct labor, machine, and other costs can be entered for the entire cell or entire plant and then prorated among completed products. In contrast, job costing requires entry of this information for each separate job through the production process.

Backflushing is the starting point for process costing. When an item is backflushed, the system calculates the cost of materials used to manufacture the product. These costs are recorded for the product or the cell, or both. Labor and machine costs can be entered individually or for the entire cell or plant and prorated across the manufactured products. Overhead costs can be similarly assigned. (See Figure 9-1.)

These costs are stored on files assigned separately to daily costs, period costs, annual costs, and even life-cycle costs. In this way, the information can be conveniently summarized and presented. Reports and inquiries show daily costs, period-to-date costs, and annual costs for products, product groups, cells, or cell groups. Information can be gathered easily — without the requirement for detailed labor-hours and machine-hours reporting — and can be presented simply and clearly.

Actual and Average Costing

The use of standard costs has come under scrutiny within many agile companies. This is not because standard costing is wrong or unhelpful, but because it often creates additional, unwarranted complexity in an otherwise simple system. Another problem is that standard costs have a great deal of waste built into their figures. Standard costs are poor measures of effectiveness because they include queue times, move times, scrap percentages, and other wasteful activities. People measuring themselves against a standard may con-

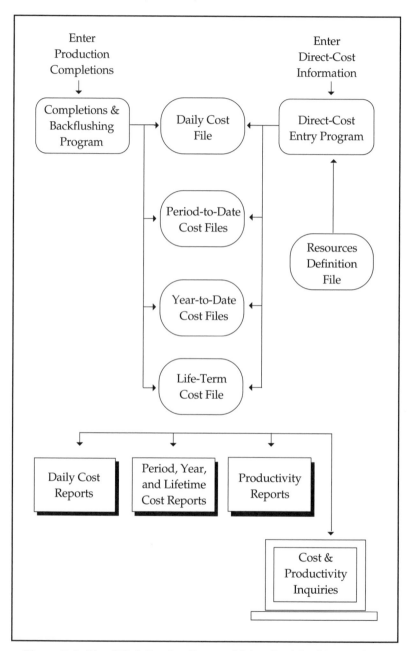

Figure 9-1. Simplified Costing Process Using Backflushing and Direct Cost Reporting

clude that they are doing a good job when, in fact, they are only achieving a standard.

One reason standard costing was initially developed was to overcome problems of manually calculating actual cost information associated with procurement and use of materials. This problem has been largely eliminated by the use of computers to perform these calculations. Using actual costing or last-in, first-out (LIFO) and first-in, first-out (FIFO) costing as the primary inventory-costing method has become quite practical.

Using average costs can be a first step toward actual costing, and most computerized inventory systems support average costing. Average costing involves calculating the average cost of materials every time a new quantity of material is brought into inventory. A company having stock of fifteen of item XYZ at $10.00 each will recalculate the item's cost when receiving another five. The cost of the additional five may have been $12.00. The new average cost will be $10.50 ($150 + $60 4 20). There are some inherent disadvantages with average costing. One disadvantage is that if transactions are not entered and calculated in exactly the same sequence the material was used, then calculating the average cost can be erratic. In fact, if issues are processed before receipts, the average cost calculations will be wrong owing to (even momentary) negative on-hand stock balances.

Most inventory-control systems have the information required to calculate actual costs. When material is received from a vendor, the actual cost of the item can be recorded; when it is issued (or backflushed), the actual cost can be used; the actual cost of finished products can be calculated directly. This information is certainly available for companies that use detailed lot traceability. Each lot of an item must be tracked separately and the cost of the lot (raw material, subassemblies, or finished products) can be retained. These individual actual costs can be accessed for use in the costing system.

There is a trade-off to be made here. If standard costing is proving to be complex and misleading, using actual costing may be better, although actual costing requires gathering and maintaining detailed cost information for products and materials. If this detailed information is required for another reason — like lot traceability — then using actual costs is feasible and practical. However, it is wrong to implement detailed tracking just for the purposes of actual costing, thereby introducing additional waste. An alternative is to use average costing that is simpler and does not require collecting additional information but does have some inherent weaknesses.

Having discussed the problems of standard costing, it is important to note that most agile manufacturers make considerable use of standard costs. The financial accounts require full absorption costing for inventory valuation; full absorption costing can often be achieved most simply through the use of standard costs. When new products are introduced, setting up standard costs that have emerged from the design process is often helpful. These standards are used through the introduction stage to assist with the attainment of the target costs established in the pre-manufacture stage.

Direct Costing

Direct costing is another textbook cost-accounting method. Although direct costing has been an option for many years, few companies have found it useful in a traditional manufacturing environment. The idea of direct costing is simple: Track only the direct costs of manufacture and do not attempt to allocate overheads to the production of a specific batch of products. There are two reasons why this approach fits well into an agile environment.

Accounting Simplification: Phase 3

The third phase of simplification requires dismantling its integration with financial accounts:

- Eliminate work orders.
  They no longer serve a useful function and retain a batch-production mentality.

- Eliminate month-end reporting.
  The tyranny of the monthly cycle can be eliminated when there is no integration with the financial accounts.

- Eliminate the integration with the financial-accounting system.
  Inventory levels are stable and there is no need for tracking detailed inventory valuation.

- Eliminate the budgeting process.
  This function has always been a futile and useless exercise in self-deception. WCM companies recognize this fact and let it disappear.

First, direct costing solves the issue of cost distortion caused by inappropriate overhead-allocation methods. The rampant cost distortion in the majority of traditional companies results in there being no clear picture of the true cost of products. Worse yet, managers and accountants often think their product costs are very accurate — perhaps correct to six decimal places — when, in fact, the numbers are misleading and dangerous to the company's profitability and survival. If overhead costs are not included in the cost calculation, there is no danger of the product costs being interpreted as anything other than direct conversion costs.

The second reason is that direct costs are more relevant than full absorption costing. Of couse, controlling the production organization using nonfinancial reporting is much better. However, if cost accounting is required for other reasons, then direct costs are at least more focused on the shop-floor tasks. People should be monitored and judged only by the elements of production over which they have control. In most companies, the production personnel have meaningful control of direct costs and very little control of overheads. Direct costing provides a clearer and more readily useful set of information than traditional absorption costing.

The argument against direct costing is that no one has a true picture of the product costs and it is, therefore, impossible to make valid business decisions. Techniques such as marginal costing analysis can be used to overcome this kind of shortcoming, as well as the use of ad-hoc analysis of costs when a decision point arises. Overhead costs must also be apportioned to the balance-sheet inventory value for financial-accounting purposes. However, doing so is not necessary for each individual product — a gross application of overheads to inventories is satisfactory for financial-accounting purposes.

The disadvantage of direct costing is that it requires direct labor reporting. Most agile manufacturers eliminate direct labor tracking and move production personnel into salaried positions. This makes the role of direct personnel significantly different because many tasks traditionally performed by middle managers become the responsibility of production operators as authority and responsibility is transferred to the shop floor and self-directed work groups. Much of this work is traditionally considered indirect. Creating a detailed tracking system to ensure that the actual tasks performed by individuals are understood and reported on a time sheet makes no sense. It is better to put everyone on salary and motivate them based upon the world-class objectives of the organization. Removing the distinction between salaried and hourly personnel is also an important team-building initiative. Therefore, the use of direct costing can be difficult in an agile company because detailed labor information is not available.

For direct costing to work well when direct labor reporting has been eliminated requires one of two options: (1) either the direct costs are limited to material costs that can be gathered easily from the backflushing and other inventory transactions, or (2) a process-costing approach is used, where actual labor costs are entered as one figure and prorated across completed products. Both approaches are valuable depending upon the individual company's needs and characteristics.

The software issues are simple. Direct costing can be incorporated into the shop-floor control software by setting the overhead rates to zero. Overhead amounts apportioned during production reporting are calculated from a table of overhead rates applicable to labor hours, machine hours, labor costs, machine costs, or component-material costs. If these rates are set to zero, then no overheads are applied. If variance reporting is still used, overhead information must be removed from the variance-report formats so that meaningless overhead variances are no longer displayed.

Activity-based Costing

Over the last five years new approaches to the allocation of overheads have been developed. The most prominent of these is the concept of activity-based costing (ABC).[3] ABC is a simple idea: If cost distortion is caused by allocating over-

heads incorrectly, then change the method of overhead allocation. Another way of looking at ABC is as an attempt to take
costs that were traditionally considered indirect overheads
and turn them into direct costs by finding a valid method of
applying those costs directly to a product or product family.

In the early years of the twentieth century when ideas
of cost accounting were being developed, overheads were a
small contribution to the total cost of a product and labor
costs were a big contribution. These days, the opposite is true.
An ABC analysis breaks down the overhead costs into the
activities that generate those costs and then determines what
drives those activities.

A "cost driver" is a factor that creates or influences
cost. For example, it is possible to calculate how much money
a company spends on engineering changes on the shop floor.
The cost is spread among many departments including
design engineering, production engineering, the drawing
office, production control, materials planning, the production
plant, and perhaps marketing and sales. If it can be determined how much time and money is spent in each department raising and processing engineering-change notes
(ECNs), then the cost of an average ECN can be calculated. If
a count is made of how many ECNs have been raised for each
product, then engineering-change costs can be assigned
directly to each product by simply multiplying the cost of an
ECN by the number of ECNs raised this month or this year.
The number of ECNs is the cost driver (the factor that creates
or influences cost); the total cost of processing ECNs is the
cost pool (the grouping of all cost elements associated with an
activity).[4]

Similarly, the majority of indirect costs can be applied
to individual products or product families using a range of
cost drivers. Cost drivers vary from one company to the next,
but often include such items as the number of purchase

orders, the number of unique parts in the bill of material (BOM), the distance moved throughout the production process, the number of vendors, and the number of customers, as well as the traditional labor hours, machine hours, and material costs.

Activity-based costing can be thought of as the same as traditional cost accounting except that overheads are applied using a wider range of allocation methods. In reality, however, the ABC approach is radically different because it forces the management accountant to examine the production process and look at where costs are created within the process. The analysis associated with ABC requires a level of understanding of the production process that few management accountants previously had. The result is a radically new approach to cost management that has become known as "activity-based management." The initial objective of the ABC designers was to improve a company's calculation of product costs. The end result is activity-based management, which is a powerful tool for perfecting the process and for continuous improvement. It can also be the technique that enables management accountants to become part of the world-class team — because they now contribute to rather than hinder progress.

When to Use ABC

There is a raging debate going on concerning the use of activity-based costing. Everyone agrees that ABC can effectively provide a better understanding of the true cost of products and product families within a manufacturing organization. Similarly, most academics and practitioners recognize that a more important aspect of ABC is the ability to fully understand the production process and where costs derive. The controversial question is whether ABC should be used for analysis only or whether it should become a part of the day-to-day cost accounting of the organization.

One school of thought is that the analysis is complex and should not be conducted very often because the information obtained does not vary significantly from one week to the next. The analysis should be used to establish a path for change and improvement and should be repeated no more than twice a year or when there is a significant change in product mix or production process. The other school of thought states that ABC is a powerful and accurate method of tracking production costs and should be used to monitor production process on an ongoing basis.

The question of simplicity and waste is important. An ABC analysis is time-consuming and adds no value to the product.[5] If the analysis is used as an ongoing part of the cost-accounting system, the company is taking on a wasteful process that replaces an inadequate and expensive costing system with one that is an improvement but is still expensive and wasteful. Eliminating the detailed cost accounting is far better than replacing it with a more sophisticated and complex method. However, an occasional ABC analysis is valuable if used for process improvement and to attain other agile manufacturing objectives.

Software Support for ABC

Several new software products available are designed to support ABC analysis.[6] Although most of these products are designed to operate on a stand-alone basis on a PC, a Macintosh, or a work station, one or two companies have integrated ABC applications into production planning and control systems that run on mid-range or mainframe systems.

These products analyze data that have been downloaded from a production and accounting system. The analysis includes the assignment of department costs to cost pools, the derivation of cost-driver rates for each driver associated with a cost pool, and the calculation of product costs using activities. Each product has its own characteristics. Some of the packages are graphical (using Windows or other graphic user interfaces) and employ charts and graphs to present the information. Others have a more engineering design approach and provide the goal-seeking capabilities required for value-engineering analysis.

The integrated products have features that allow the user to employ ABC as an ongoing cost-accounting system, drawing data from the general ledger and the production-reporting transaction to derive the cost-driver quantities and applying activity-based allocations to production-completion quantities. Some ABC packages are available for individual purchase, others must be linked to a larger system, and some can be used only when employing the services of the consulting organization that owns the software.

Most ABC packages are reasonably priced and some offer attractive and easy-to-use features. Having said that, the author adds that calculations required for an ABC analysis are not very complex. A company can easily write its own customized analysis system using a spreadsheet or data-base tool. The author has successfully used simple spreadsheets to develop ABC analysis. The use of comprehensive spreadsheets such as Lotus 1-2-3, Quattro-Pro, or Excel, or data-base tools like Microsoft Access, dBase, or Paradox enable sophisticated, graphics-based systems to be created that precisely meet a company's needs.

Integrating ABC with a Production System

Integrating ABC into a production planning and control system, when required, is best done though a process-costing approach. Process costing is concerned not with completion of an individual batch of product but with the development of costs over time. ABC is also concerned with how costs are driven over time rather than with the costs of an individual production run. If a company is still using work orders, a useful approach is to extract costing information from the work order and place it into the process-costing files for ABC purposes. This method has the added advantage that there is a single set of costing files for both cellular and work order-based manufacturing. As the company changes from work orders to cellular manufacturing, costing information does not require modification.

A successful approach is to have a series of files containing the costing information. One file will contain daily detailed information, another month-to-date information, and a third year-to-date summaries. These files contain information about completion quantities, hours worked, standard costs, standard hours, and a large number of overhead cost elements. Companies commonly use fifteen to 20 cost drivers when calculating activity-based costs, and an organization might even use as many as 50. However, having too many cost drivers or cost elements is not sensible — because the more there are, the more complex and confusing the system becomes. The most useful approach is to have just a few cost drivers providing approximately 80 percent allocation of overhead costs; the remaining 20 percent can be allocated in a traditional way. The additional complexity required to capture the detailed information for the remaining 20 percent of costs is counterproductive.

When production completions are reported, either through the backflush of a production cell or a work-order completion transaction, the actual or standard material costs are carried into the process-costing files. If necessary, labor costs can be assigned or prorated into the product's actual costs. The application of overhead costs using ABC is often not done day-by-day but is posted periodically because the driver-quantity information usually is not immediately available.

There are two ways of assigning ongoing ABC costs to product. One way is to assign a cost-driver rate to each cost driver. For example, the driver rate for an engineering-change note may be $550. Every time an ECN is raised for a product, $550 can be assigned to a cost element on the process-cost files. The cost-driver rate is predetermined during the analysis stage of the ABC project. A mechanism must be created to capture the cost-driver information. In this example, the number of ECNs raised for each product must be extracted from the engineering data system and used to drive the allocation of cost. Some cost drivers are not readily available on the computer systems and are entered manually.

The second way to assign activity-based costs is to have a program that reads details from the general ledger and captures the period's expenses for each relevant department. A table is established that assigns the amount of money in each departmental account to one or more cost drivers. For example, personnel from the production-engineering department might spend 40 percent of their time processing ECNs. Therefore, the table will have a 40 percent factor for the assignment of departmental costs to the ECN cost driver. This percentage assignment of costs from departments to cost drivers is a part of the ABC analysis procedures. When all departments have been processed, the total amount of money in the cost pool for each driver can be prorated to the individual products. Product costs are then calculated according to the activity-cost-driver quantities used for each product.

Both approaches must be contained in the ABC system because some cost drivers will require obtaining the detailed departmental costs each period. Others are better handled using a standard cost-driver rate that changes only when a new analysis is completed.

Value-Added Analysis

A potent tool of agile manufacturing is the value-added analysis. Value-added analysis determines which activities add value to the product and which do not. A value-added activity is one that transforms or shapes a product or service toward that which satisfies the customer's real and perceived wants and needs.[7] A non-value-added activity is defined as waste and should be eliminated as soon as possible. A typical Western company has a production process that is around 5 percent value-added; the remaining 95 percent is waste.

Value-added analysis can be completed in two different ways: It can be a part of the company's industrial-engineering analysis or a part of the management-accounting procedures. If it is conducted by production or industrial engineers, then it will be a part of the analysis of production processes and material flow. As the engineers study the production process with a view to waste elimination and continuous improvement, they will become very knowledgable of the entire process and can determine which activities add value and which do not. This analysis will reveal potential candidates for elimination as a part of the company's various improvement projects.

Value-added analysis completed by management-accounting people is usually done as a part of the ABC analysis. An ABC analysis provides a thorough understanding of the production process (and other processes within the organization); steps within the processes can be designated value-added or non-value-added. In practice, it is better to have the four categories presented in Table 9-2.

This four-part classification enables the areas of greatest potential to be highlighted and dealt with first. In addition, it makes value-added analysis acceptable to people within the organization who might be intimidated with their entire job function being classified as "waste." To classify it as a non-value-added support function tends to sugar the pill and reduce antagonism and fear. There are no hard and fast rules relating to the classification of activities. An activity classified as "support" in an early analysis can be reclassified as "waste" as agile manufacturing progresses.

Software support for value-added analysis is not complex. It is useful for any ABC systems to contain value-added fields and to be able to present the cost information classified under the four headings. Alternatively, the production routings can be used to analyze value-added activities. Individual job steps within the routing can be designated with a value-added code and reports written showing the value-added and non-value-added activities within the production process. These reports can be produced from a history of actual production completions during the previous week or month, or the reports can be produced using the master schedule of future production. This routing-analysis approach requires production routings to be complete, current, and correct.

Costing for Cellular Manufacturing

The ideal approach to cost accounting for cellular manufacturing is to eliminate it and to control the production process with nonfinancial reporting. If cost tracking is required, then simplified cost-accounting methods should be employed. Cellular manufacturing lends itself to a process-style costing, and the completions reporting is the starting point.

Adds Value	Value Added	directly adds value to the product or service from the customer's point of view
Strategic	No Value Added	non-value-added activity (like training) that is essential to the organization's long-term well-being
Support	No Value Added	non-value-added activity that (currently) provides a service to maintain the operation
Waste	No Value Added	adds no value to the product or service and has no strategic or support function

Table 9-2. A Four-Part Classification Table

The completions-reporting program creates a transaction to book the finished product (or subassembly) into inventory, updates the production schedule, and backflushes the components to relieve inventory and book the parts to production. Within this same process, the program can also create a costing record showing the completion of that item in the cell. (Refer back to Figure 9-1.) This costing record will contain the completion quantity information, the standard cost of the completed product, the standard production hours, and the actual material costs derived from the back-flushing process.

If labor reporting is required, a program can be used to enter labor hours or costs for the cell and prorate those costs across the product completed in the cell. In a similar way, machine hours and machine costs can be apportioned. This approach can be expanded to have a table of resources (including labor and machine hours) so that the costs of other resources can be applied to the completed products. This method is a simple first step toward the reporting of activity-based costs. Traditional overheads can also be calculated when the labor, machine, and material costs are entered. The resource table contains an overhead percentage that is used in the calculation of overhead costs. Each resource cost and its associated overhead can be assigned to one of the product cost elements (such as material, labor, overhead, and outside process). The resource table contains the cost element associated with each resource.

Data contained on the cost files can be reported in a number of different ways providing cost information for products, for cells, and for combinations of products and cells. While the information can be shown for specific dates, month-to-date and year-to-date information is more useful because it shows trends as the costs change over time.

This approach to cost accounting for cellular manufacturing is simple and requires a minimum number of transactions. However, many companies using cellular manufacturing find that even this level of detail is not required. Inventory-control systems contain information about the products completed during a period, and the general ledger contains total information relating to costs (both direct and indirect). Simple reports can be written that extract cost information from the general ledger and match this to the completed products. These reports can provide powerful information about cost trends and production efficiencies without the need for additional cost-collecting activities.

Simplifying Accounts Payable

The key to simplifying the accounts-payable (AP) process is to eliminate the non-value-added activities, which in most companies is the entire process. Implementation of JIT manufacturing puts considerable strain on a traditional accounts-payable system because there is so much more material-delivery activity. A traditional system requires a receiving transaction for each receipt and an invoice for each receipt. The accounts-payable personnel enter the invoice information and match it to the receiving information to ensure accuracy and to calculate purchase-price variances. These price variances are posted to the general ledger and used for measuring the performance of the purchasing department's personnel. This kind of procedure can only work when a company has relatively few material receipts. As daily JIT delivery becomes the normal method of operation, the accounts-payable system becomes unwieldy.

These traditional procedures were devised for companies that have adversarial relationships with their vendors. The purpose of receiving invoices and matching them is to ensure that the vendor is not charging incorrectly and to check that the inventory transaction has been processed properly. As discussed in Chapter Seven, the agile approach in these issues is to create close, mutually beneficial relationships with the vendors, relationships that eliminate the need for these kinds of checks and balances.

These accounts-payable procedures can be eliminated in stages. The first step is to consolidate the receipt information and receive a single invoice for multiple deliveries. This step cuts down the amount of work required within the accounts-payable department. The next step is to eliminate the need for an invoice by creating the invoice records in the accounts-payable system when the receipt occurs. This action represents a significant savings in the accounts-payable department because invoices no longer have to be entered by hand. The vendor also saves a considerable amount of non-value-added work printing and mailing invoices.

The third stage is to eliminate the receiving transaction, by processing the receipt as part of the backflushing procedure when the product is completed. Because the component materials used in producing the product must have been received from the vendor, a receiving transaction can be created. If a receipt has occurred, then an invoice record must be created.

The final stage is to eliminate the process of selecting and printing checks by using electronic funds transfer (EFT), which also can be triggered by the backflushing process. A JIT payment can be made every time a product is completed. However, it is more common for the EFT to be a consolidated payment made daily, weekly, or monthly to the vendor based upon all receipt transactions created by the backflushing process.

Simplifying Accounts Receivable

The accounts-receivable (AR) process is the mirror image of the accounts-payable process. The key to simplification is to create close and mutually beneficial relationships with customers. Within an atmosphere of trust and mutual dependence, it is possible to gradually introduce the changes discussed in the previous section relating to accounts-payable activities.

For most organizations, all operations of an AR department are non-value-added. Typical activities of the AR department are credit checking, creating invoices, receiving payments, and expediting delinquent customers through a variety of dunning notices, telephone calls, and legal action. While these activities will always be necessary to some extent because a company has new customers with whom a close

---

Accounting Simplification: Phase 4

The final phase of simplification:

- Eliminate traditional cost accounting.
  With traditional methods no longer relevant, turn to nonfinancial methods of control and monitoring.

- Backflush to accounts payable.
  Use backflushing to eliminate the receiving and invoicing process.

- Use electronic funds transfer.
  Eliminate checks and receipts and go to just-in-time payments.

relationship has not yet been established or casual customers who make a single purchase, the objective of an agile manufacturer is to create customer relationships that eliminate the need for traditional accounts-receivable activities.

When a company first begins to implement agile methods, the AR department will be designated as a non-value-added support department. Changes in receivables are not the first priority. However, as agile concepts begin to take root within the organization, the techniques of total quality, short cycle time, waste elimination, and synchronization become as applicable in the clerical areas of the company as on the shop floor.[8]

Multicurrency

As manufacturing and distribution marketplaces become more international, the need for multicurrency software increases. There are four aspects to multicurrency systems — currency tables, customer order entry, procurement, and general-ledger activities.

Currency conversion tables, the heart of a multicurrency system, need to provide considerable flexibility for present and future needs of the organization. It must be possible to work in many currencies and to have many different types of exchange rates within each currency. The exchange-rate types enable the user to apply different exchange rates for different occasions and needs. The exchange-rate types will include a plan rate used for budgeting or planning purposes, an intercompany rate used for internal transactions (indeed, many multinational companies have more than one intercompany rate and update them quarterly), a spot rate, an average rate, rates used for specific customers, rates used for currency bought forward on the money markets, and so forth.

For each combination of currency and exchange-rate type the system allows the user to enter the exchange-rate value. It is often useful to have the choice of whether the exchange rate is expressed as a multiplier or a divisor ($1.51 U.S. to the British pound, or £ 0.66225 to the U.S. dollar) and to have a choice of decimal places for the currency. Many currencies that have experienced considerable inflation do not require two decimal places in the accounting systems. There is also a field-size issue here because the financial amounts within the system must be able to accommodate currencies like dollars, deutsche marks, or pounds, where two decimal places and seven places before the decimal is enough for most standard transactions. They must also accommodate currencies like the peso or lira requiring no decimal places but needing ten or eleven places before the decimal. The exchange-rate table needs to show this decimal information, and all screens and reports within the system must adjust the use of the financial-amount fields according to the field size needed by that currency.

The exchange-rate table also has date effectivity. The rate entered applies to the currency code and exchange-rate type for the range of dates entered by the user. This allows multiple rates to be held on file and a history of the exchange-rate changes to be maintained. The system takes account of the date when the exchange rate is selected for use. When an invoice is created in the accounts-receivable system, the exchange rate selected will be the rate applicable at the time the invoice is created. Some systems require only one rate to be effective at any one time; other systems are more forgiving.

Key transactions with the system must have currency codes and exchange-rate types assigned to them. Each vendor (as does each customer) has a standard currency code and exchange-rate type. These indicators, set up on the vendor master file and the customer master file, are used in all transactions relating to these entities unless they are overridden manually.

The currency conversion tables also require bank-account information. Each combination of currency code and exchange-rate type has a bank account (or other financial institution) associated with it. This account is used when payments are made and receipts deposited. Not every combination of currency code and exchange-rate type requires a different bank account; the same bank account may be shared by several combinations. In addition, the exchange-rate tables in some systems contain accounts-payable and accounts-receivable control accounts so that their balances can be recorded separately on the general ledger.

The tables also contain a gain-and-loss account and revaluation account for each currency (or combination of currency and type). These accounts are used to post realized and unrealized gains and losses for the currency. A realized gain or loss is posted when a cash payment is made or received, when the exchange rate has changed between the time the invoice was created or entered and the payment occurred, and the effect of the exchange-rate change has been felt within the accounts. An unrealized gain or loss is posted when the inherent value of a payable or receivable item changes due to an exchange-rate change but where the item is still an outstanding payment or receipt.

Customer Order Entry and Multicurrency

Customer order-entry and accounts-receivable systems require specific features to meet the needs of multicurrency operations. The currency code and exchange-rate type for a customer (or a ship-to location) is pulled into the order-entry process from the customer master file. This information enables the system to select the appropriate exchange rate for the sales order. All sales-order transactions are then shown in the "foreign" currency despite the fact that such factors as price and costs are retained on file in "base" currency. The information is usually held on file in base currency together with the exchange rate so that amounts can be converted to foreign currency as required.

Although prices and costs are held in base currency, it is often useful to hold customer-specific pricing in foreign currency. Customer-specific pricing is a feature many systems have that enables product prices to be established for a specific customer. The customer-specific price file may also contain special discounts, part-number conversions, and certification codes. Holding these customer-specific prices in the currency of the customer makes sense because the agreement between company and customer will have set the prices in the currency in which the transactions occur. To recalculate these prices into base currency would be a fruitless task.

All "external" documents relating to the sales orders are printed in foreign currency and all "internal" documents are printed in base currency. On-line inquiries within the system are in base or foreign currency according to the use of the inquiry, but will have the ability to "toggle" between foreign and base currency as required. Reports that contain information from many orders and are in multiple currencies are primarily printed in base currency. However, they have an option to allow printing them in foreign currency. (In this case, the sort sequence is usually currency code first so that totals and subtotals are meaningful and do not contain mixed currencies.)

Once the products are shipped to the customer, an invoice is created in the usual way. The invoice is printed in foreign currency because that is the currency in which the company expects to receive payment. Payment, when received, will usually be in the currency of the invoice; and the accounts-receivable system accepts payment in that currency. Most systems allow the payment currency to be overridden so that the customer can pay the invoice in a different currency from that of the invoice. The system then has to calculate a double currency conversion so that the invoice can be updated in the invoice currency and the general ledger updated in the base currency from a payment made in a third currency.

Procurement and Multicurrency

Procurement processes are the same as the customer sales order. Purchase orders are created in the currency of the vendor although the user can override this currency code manually if required. External documents are all printed in the foreign currency while the internal documents are printed in base currency. Costs are all held in base currency except for vendor-specific costs that are held in the vendor's primary currency.

Generally, the invoice received from the vendor will be in the same currency as the purchase order, but most systems allow the invoice to be posted in a third currency. Some systems have a feature to allow the material to be received in a third (or fourth) currency so that the average cost calculation and purchase-price variances can take account of the exchange-rate changes prior to the invoice being received. All transactions within the system are posted in base currency. The currency code, exchange-rate type, and rate itself are held on the file so that the item can be presented on reports and inquiries in either base or foreign currency.

When payment is made using electronic funds transfer and no invoice is received from the vendor, the currency will have been agreed upon ahead of time by supplier and customer. Often not only the currency is decided but also the exchange rate itself so that the commitments between the two organizations are arranged clearly. Within the United States, EFT in any currency other than U.S. dollars is unusual; even Canadian companies use U.S. dollars when doing EFT. However, in Europe it is common to find EFTs completed in different currencies for different vendors.

Currency Revaluation

An important aspect of multicurrency features is the revaluation processes. These programs revalue accounts-payable and accounts-receivable invoices that are outstanding on the system. The purpose is to comply with accounting standards which state that assets and liabilities on the balance sheet must be shown at the value on the balance-sheet date. Revaluation programs read through the AP and AR invoices to calculate their latest values. The new values are posted to the general ledger so that the AP and AR control accounts are "accurate," and net changes for each currency are posted to a revaluation or unrealized gains-and-losses expense account.

Revaluation programs usually have a "report only" feature so that the revaluation process can be completed and the report printed without the ledger being changed. This capacity allows accounting personnel to review the effect of revaluation changes before the financial postings are completed. There is often a choice of how the revaluations are posted. Some companies wish to retain transactions at their original rate until they are paid and a realized gain or loss taken. In this case, the revaluation programs do not directly affect the transactions but post reversing accruals to the general ledger. This method enables the balance sheet to be valued correctly for reporting purposes — although the effect of the revaluation is reversed out the day after the balance-sheet date. Other companies are happy to have the revaluation posted permanently to the general ledger but wish to post the changes to separate accounts so that unrealized gains and losses can be reversed out when the realized gain or loss is posted. The revaluation programs use a posting table to select which accounts are posted during the revaluation process; this system provides the flexibility required by different companies.

General Ledger and Multicurrency

Issues relating to the general ledger are significant only when a company has multiple divisions that operate in different base currencies. The consolidation process is made considerably more complex by the multicurrency aspect of the procedure. Each division's ledgers must be converted to the corporate base currency, and offsetting gains and losses between divisions must be balanced out. In addition, different accounting regulations apply to different countries. These differences must be eliminated from the ledgers so that the consolidated reports adhere to the regulations of the home country.

The consolidation of multicurrency ledgers is a complex process requiring a considerable amount of calculation and posting and a clear understanding of the various international accounting procedures. The process is not a clear-cut science because there are many areas where experience and judgment must be used to determine the appropriate actions. Consequently, it is not possible to create a computer system that automatically handles all the eventualities and eccentricities of the process. The best approach is to have a feature in the accounting system that converts the various ledgers into the home country's base currency and places the information on separate files. This procedure allows the various company divisions to continue working while the consolidation process is being conducted. It also gives the people doing the consolidation the flexibility to make their adjustments and "what-if" analyses on a separate set of data without touching the "live" files.

Standard Cost Buildup

Most agile companies use standard costs. They do not do variance reporting, they do not track actual against standard, they do not judge people or departments according to their adherence to standard. However, they do use standard costs for inventory valuation, establishing interplant transfer prices, and simplifying cost calculations.[9] The development of standard costs is much simplified if the software contains a standard-cost-buildup facility.

A standard-cost-buildup system explodes the bill of materials for each selected product and accesses the production routings. From this data, and from the information relating to overhead rates, labor, and machine costs, the system calculates the standard costs for the product and for subassemblies at each level through the BOMs. Most cost-buildup applications allow the user to create a cost buildup for a single product, a range of products, or the entire file of products. Results of the buildup calculations can be held on a temporary "what-if" file so that reports and analyses can be performed before standard costs are applied to the live data base. This system enables the accountants to gradually develop new standard costs (perhaps for the following year) and then apply them to the live system at the appropriate time.

Most cost-accounting and inventory-control systems have a limited number of cost elements for a product. These elements will include material, labor, machine, fixed overhead, variable overhead, outside processing costs, and perhaps one or two other categories. Cost-buildup systems often allow the user to establish a larger number of cost elements for analysis and calculation purposes and then to "roll" those detailed costs into the standard cost elements in the final stage of the calculation. Such a procedure provides the best of both worlds because the accountants preparing the standard costs require considerably more detail when developing standard costs. However, the day-to-day user of the standards does not want the confusion of multiple cost elements. This approach allows the accountant to use an ABC-like approach to the cost-buildup process and yet retain a traditional standard cost for the ongoing user.

Some software systems allow separate BOMs for cost-buildup purposes. The idea is that the accountants developing the standard costs need the flexibility to change the BOMs and routings for standard costing purposes without affecting the bills and routings used in the live system. While this feature is often attractive to the accountants, in fact, it is dangerous nonsense. One problem of using standard costing is that the concepts inherent within the standards are divorced from the practical activities on the shop floor. This factor is part of the divisiveness that exists in most companies between management accountants and production people. An agile company actively breaks down this barrier and brings accounting people into the team with the production, engineering, procurement, sales, marketing, and design personnel. Accountants developing the standard costs should use the same BOMs as the shop-floor, engineering, and design people.

Many cost-buildup systems support incremental costs and elemental costing. With incremental costing, the total standard cost of a subassembly (including all cost elements) is applied to the material cost element of subassembly or finished product the next level up the bill of material. With elemental costing, the individual cost elements of lower-level subassemblies are added into the corresponding cost elements of the next higher level. Also, the breakdown between material costs, labor costs, machine costs, and other costs is retained up through the bill to the finished product.

Companies using cellular manufacturing methods strive to "flatten" their BOMs, ideally to a single level, thus simplifying the calculation of standard costs. If the product goes from raw material to finished product in one step on one cell, then standard cost can be easily calculated. However, many companies have a two- or three-step production process with the product passing through more than one production cell. They do not want the expense and complexity of reporting subassembly manufacture because they never inventory these interim production assemblies — they merely pass from one cell to the next. Should it be necessary, the standard cost buildup can take account of this methodology and calculate the total standard cost for the final assembly. It can also calculate the interim standard costs for a partially completed product within one of the cells.

Summary

Traditional accounting systems, particularly management-accounting systems, are unsuitable and harmful to agile manufacturers. These systems are expensive and wasteful to operate, provide misleading and potentially damaging information, lead people to do the wrong things, and do not lend themselves to continuous improvement.

While many agile companies have found considerable problems with their management-accounting systems, they often find it difficult to remove or modify them because they are so much a part of the company's infrastructure. This fact is particularly relevant when the organization is a division of a larger corporation that mandates management-accounting reporting and may not have a total commitment to agile concepts.

The best approach to management-accounting systems is to largely eliminate them from the day-to-day operation of the production and distribution plant and to run the plant using nonfinancial reporting and control. Financial information, when required, can be obtained from the financial-accounting system or from simple reports derived from other parts of the planning and control system. If the management-accounting systems cannot be eliminated altogether, steps should be taken to eliminate their undesirable aspects. This would include the following:

- Eliminate detailed labor reporting.
- Eliminate variance analysis.
- Reduce cost-center reporting through cellular manufacturing.
- Eliminate work-in-process inventory tracking through the use of backflushing.
- Eliminate work orders and use simplified rate-scheduling or no scheduling.
- Eliminate monthly reporting, integration with financial accounts, and detailed budgeting.
- Simplify the accounts-payable and accounts-receivable systems by close partnership with customers and suppliers.

Some techniques that are important to agile manufacturers are:

- process-style costing (instead of job costing)
- the use of direct and actual costing to eliminate cost-distortion
- activity-based costing to establish more accurate product cost and to provide a tool for understanding and improving production (and other) processes

Common to all world-class methods, simplification and waste elimination are the touchstones for accounting systems.

———

CHAPTER TEN

Product Design

With their emphasis on fast time-to-market, design quality, and design for manufacturability, agile manufacturers take a radically different approach to product design and development from that of a conventional manufacturing company. The keys to the use of computer systems in the area of agile product design are: (1) selecting software that supports a world-class approach and (2) integrating the various systems used in the product-design process. This integration must occur not only within the product-design software, but also within production and inventory planning systems, engineering systems, and — if appropriate — the sales and marketing systems. Many companies use product-design software of one sort or another. However, few companies gain the full benefits that can accrue when an excellent computer-aided design is integrated with computer-aided process planning, automated production equipment, and the manufacturing planning and control systems. A world-class manufacturer — concerned to eliminate waste, cut total cycle times, and provide greater flexibility to meet customer demands — integrates the product-design processes with production processes. This fast response, design-to-order approach can be a powerful competitive edge.

Two key aspects in these new approaches to product design are computer-aided design (CAD) and concurrent engineering. CAD includes the use of software that is intended to simplify the design process, improve the quality of the designs, and automate the production of drawings. In addition, CAD systems can be interfaced to computer-aided manufacturing (CAM) equipment like numerically controlled (NC) machines and robots. Concurrent engineering includes a number of techniques whose collective purpose is to design the product right the first time. This approach requires cross-functional teams of people working together to design all aspects of the product — from matching customer needs to establishing target product costs to developing manufacturing methods. The overlapping of tasks and the team approach to design enable agile companies to create new products (and major product enhancements) faster and better than traditional companies.

Time-to-Market and Teamwork

One of the few things we know about the future is that it will differ from today. Everything is changing at bewildering speed — including the rate of change. Companies that prosper as we approach the 21st century will be those that are quick on their feet and able to adapt to changing technologies, changing markets, and changing business needs. One aspect is the introduction of new innovative products. The necessity for companies to bring new products to market quickly and effectively is already apparent and will become more critical over the next few years. In most industries, successful companies will be those that can quickly bring to market products that meet — or exceed — customer expectations and are competitive with the best in the world.

The trick is to develop a design process that provides scope for imagination and innovation, takes the product from concept to manufacture very quickly, and designs the product right the first time. Included in these requirements is the integration of marketing, design engineering, cost accounting, production engineering, quality, sales and distribution, and post-sales customer service and support. Thus, a sophisticated team approach to managing the design process is essential, necessitating the elimination of departmental boundaries and solidifying a common understanding of the organization's design goals and objectives.

Traditional companies design products linearly. Design engineering is done first, followed by production engineering, followed by costing and pricing, followed by sales, and so forth. There is very little interaction between the clearly defined departments, each of which has different tasks, responsibilities, and outlook. Indeed, many companies experience active distrust and rivalry between the departments that result in additional waste and conflict. On the other hand, agile companies seek to compress the task of designing innovative new products by combining the various activities into focused, cross-functional teams. Doing so has a double benefit: Not only is the design completed more quickly, but also the quality of the design is greatly enhanced because design teams are able to take account of every aspect of the design process within a single design activity. Products are designed for manufacturability because manufacturing personnel are included in the design process from the beginning. The product meets the needs of customers because sales and marketing personnel are incorporated into the design process. The management accountants contribute to the design process through the use of new accounting techniques like target costing, value engineering, and life-cycle costing. This approach to design, which is not easily accomplished, provides agile manufacturers with innovative new products more quickly and with greater quality and manufacturability than the traditional approach.

Design for Quality, Cost, and Manufacturability

An agile manufacturer's definition of quality is broad. It includes the idea that quality requires meeting customer needs and expectations as well as making the product according to specification with consistent high quality. The concurrent-engineering process uses techniques that enable designers and marketing personnel to formally match expressed and implied customer needs to the specific product features. One technique used is quality function deployment (QFD), which employs a visual chart to catalog and prioritize customer needs and match them to various features of the product throughout the design process. QFD charts (sometimes known as the "house of quality" because the charts are shaped like house plans) also facilitate a structured approach to competitive analysis by comparing current and future products from competitors with the features and functions of the products being designed.

The design process also includes detailed attention to production quality. The active involvement of quality engineers, production engineers, and shop-floor personnel as a part of the design team ensures that the product being designed can be manufactured to the highest quality. Techniques such as fail-safe, modularity, and commonality can be included in the design phase of a product much more effectively in contrast to the traditional approach of re-engineering products and processes after production begins.

The idea behind designing for manufacturability is to design a product that does not have to be modified to manufacture it effectively. Most traditional companies find that an elaborate pilot production is required to iron out the bugs in the product design. This pilot process allows production engineers to modify product design so that products can be manufactured more easily and at lower cost. In addition to pilot production, a considerable number of modifications and engineering changes are required within the first few months of production as the process is improved and standardized. This expensive, time-consuming process can be eliminated if the product is designed with manufacturability in mind from the outset.

Design for manufacturability requires the inclusion of manufacturing and production specialists on the product-design teams and the use of techniques like group technology, modular design, and part and process commonality. Production people will also ensure that other important aspects of agile manufacturing are taken into account through the design process. These aspects include fast changeover of machines and tooling, small batch sizes, commonality of tooling, cellular manufacturing, product and process simplification, and differentiation of the product in the later stages of manufacture. The objective is to design a product that can be incorporated immediately into production without the need for lengthy pilot production and a lot of engineering changes in the early months. As product life cycles get shorter and shorter, it is increasingly important not only that the design process is compressed but also that products can be immediately manufactured according to the required cost, quality, and production lead time.

Computer-Aided Design

Computer-aided design has already revolutionized many manufacturing organizations; most Western manufacturers utilize some CAD in their facilities. Although CAD is common to most of these new techniques, only a few companies use CAD to its full potential. CAD is a revolutionary new way of designing products making use of computing power to improve design, improve quality, speed up the design process, and help the design team to communicate effectively. Unfortunately, many design managers think of CAD merely as being automated drafting. They continue to approach new product design in a traditional way by using CAD systems at the tail end of the process for printing drawings. Such limited usage of CAD misses the potential use of computer systems for product design.

CAD Benefits

The use of computer-aided design is very much in line with the ideas of agile manufacturing. When implemented and used well, CAD supports the use of cross-functional teams, improves quality, eliminates waste from the design process, facilitates a right-the-first-time mentality, reduces design cycle time, facilitates product flexibility, and encourages innovation.

Ian Williamson of Peat Marwick in London identifies the following CAD benefits:[1]

- *Low price.* CAD can optimize material use, reduce scrap, encourage the use of standardized parts, and standardize production processes and tooling that lead to reduced inventory and low-cost production.
- *Product performance.* CAD can be used to analyze and optimize design and function before production.
- *High quality.* CAD can be used to analyze stress, part interaction, the effects of wear and failure, and other technical aspects of a product.

- *Rapid response to market changes.* Standard parts and accurate manufacturing data speed the development and production of new products.
- *Customization.* CAD facilitates rapid modification of existing designs and manufacturing data.
- *Innovation.* CAD allows greater flexibility of design. More options can be examined in the same time frame, exotic materials can be tested, and extensive visualization of products can be undertaken.

Hardware and Software

CAD systems are available on a wide range of computer equipment. Owing to CAD's graphical nature, the computers require considerable power to process and present the information. Early CAD systems were written for large mainframe machines. However, the use of minicomputers that brought powerful machines right into the design shop opened up the widespread use of CAD. More recently, work stations and networked PCs have enabled smaller companies to utilize these new technologies. New kinds of hardware have grown up around the CAD needs of design organizations. These new developments include various kinds of large display screens and monitors, specialized printers and plotters, and novel methods of data input including light pens, scanners, and specialized mouse devices.

CAD software varies enormously. Some programs are powerful and comprehensive, while others provide elementary features. Some are used primarily for drafting, others for design analysis, and still others focus on graphical display. The latter are used when the product's aesthetic appearance is of paramount importance and the design requires artistic as well as engineering input. In addition to generic CAD systems intended for a wide range of companies and uses, a multitude of CAD systems are designed for the needs of specific industries or products. For example, the design of printed circuit boards requires specific analysis that can be incorporated into the CAD system. In a wider context, architectural CAD systems containing building regulations can alert the architect when a design violates codes.

The four aspects of a CAD system are:

1. design analysis
2. drawing
3. manufacturing documentation
4. data exchange

While not all CAD systems address all four aspects, most have at least an element of each. CAD design analysis replaces the traditional approach to modeling. Common to traditional product design is the manufacture of models or prototypes of future designs. These models enable designers to better visualize the item, test its characteristics, and tangibly communicate the design to other people within the organization. Good CAD systems provide excellent three-dimensional (3-D) representations of a product during its design. These drawings help the designer visualize the new product and facilitate communications within the design team. Because the 3-D models can be created rather quickly, any ideas and suggestions of the design team (and other interested parties like potential customers) can be investigated and reviewed more easily. The result is better-designed products that match customer requirements.

In addition to visualization and modeling, CAD systems can also perform other sophisticated and specialized analysis. Finite-element analysis is commonly available within CAD systems. This analysis tool automatically breaks down the design model into smaller elements with standard geometry. The elements can be analyzed using known characteristics of those geometric shapes. Mechanical characteristics can be analyzed using stress analysis and a knowledge of the materials from which the items will be made. The topography of printed circuit boards or integrated circuits can be analyzed to ensure their integrity, a complex and time-consuming task to do manually. Specialized CAD systems contain analysis tools that are specific to the kind of product being made and the design process being used.

Three kinds of 3-D visualization are supported by CAD systems. The first, *wireframe modeling*, is the simplest, easiest to program, and least costly on computer resources. Wireframe modeling visualizes the item being designed and creates a 3-D picture in the form of a series of lines tracing the shape of the item's surfaces. Models created this way look like they are manufactured out of shaped wires. While it is commonly used, wireframe modeling has some important limitations.

A better kind of modeling, *surface modeling*, creates the model in a similar way to wireframe modeling except that it fills in and shades the surfaces. This system gives a more realistic view of the item and is particularly useful when designing mechanical items in which the form of a particular surface is critical to the design.

The best kind of modeling is *solid modeling*. Although it requires considerably more computing power and is slower, it provides a far more realistic view of the item being designed. It also enables design changes to be made visually on the 3-D model, instead of having to create a new model each time from a two-dimensional (2-D) drawing or layout. In addition to designing the product itself, solid modeling is also helpful with process design. For example, the individual steps required to manufacture a machined part can be modeled on the computer. This capability assists enormously with the quest to design products that are easy and foolproof to manufacture.

Drafting and Manufacturing Documentation

The use of CAD systems for drafting is widespread and well understood. Once the design process is completed, communicating the design to a range of people throughout the organization is necessary. This procedure often requires 2-D drawings and manufacturing documentation such as a bill of materials, a production routing, parts lists, quality assurance (QA) criteria, inspection procedures, programs for numerically controlled machines, and technical manuals.

All CAD systems have drafting capability (and some specialize in drafting). In addition, most CAD systems support manufacturing documentation in one form or another. The more sophisticated CAD systems have libraries of technical information about components and materials and are able to recommend specific components to the designer through the use of group technology. This feature can be helpful when standardization is important. The CAD system seeks out components, subassemblies, and processes that have similar group technologies to the items being newly designed. Instead of creating entirely new components, processes, and procedures, the designer can use preexisting elements in his/her design, resulting in less complexity, lower risk, and substantial cost savings.

Graphic output of CAD systems is not limited to engineering drawings and representations. Many CAD systems incorporate desktop publishing features so that items such as sales literature, technical specifications, and high-quality illustrations can be printed. These graphics usually require specialized printing and plotting equipment using multiple color and graphic-design techniques.

Data Exchange

The majority of companies use CAD on a stand-alone basis. The potential benefit to agile manufacturers from computer-aided design and computer-aided manufacturing (CAD/CAM) technology is enormous. The three kinds of data-exchange requirements are (1) production-planning and inventory-planning data, (2) engineering data, and (3) programs for programmable machine tools.

Production-planning and inventory-planning data include parts lists, effectivity dates, product-family information, and production routings. This information can be used by MRP or other material planning systems to schedule the purchase of materials from vendors and to schedule production.

Engineering data include bills of materials and production processes, test procedures, inspection criteria, and other engineering characteristics such as safety codes and shelf life.

Programs for numerically controlled machines are used when the organization uses NC machines in production and where the CAD system supports the programming of NC machines. Instead of a technician reading the engineering information and manually creating programs for NC machines, the CAD system can automatically create these programs from the specifications contained within the design. A major problem (discussed in detail in Chapter Eleven) is the lack of standard methods for communicating this kind of data. While a number of initiatives are being offered to create industrywide international standards for data communications, these have yet to become widely accepted.

Increasing use is being made of electronic data interchange (EDI) for the transfer of design and process information within a company. The ANSI X12 standards (as well as the European standards) contain record layouts for the transfer of design and drawing information and can be used when transferring information between different plant locations within an organization. EDI can also transfer material and production planning information from the CAD systems into the production planning and control systems.

Computer-aided Process Planning

The use of computer-aided design systems is widespread. The use of computer-aided process planning (CAPP) is less popular because the systems are more complex to introduce and use. Of course, the design of the production process is as important as the design of the product itself, and integration of product design with process design using cross-functional teams is central to an agile approach to design.

The design of production processes varies enormously from one company to the next according to the products, production technologies, and market needs. However, three generic problems must be solved when designing a production process in an agile manufacturing environment: standardization, speed, and quality. Many aspects of the speed issue (or time-to-market) are dealt with by the use of cross-functional teams working concurrently to create the product and process design. Other aspects, such as elimination of waste in the product-development process, can sometimes be greatly assisted by the use of a well-designed CAPP system.

It is in the area of standardization and quality that a CAPP system can be most helpful. Standardization reduces complexity, reduces risk, and shortens the learning curve during the introduction of a new product. Objectives of a CAPP system include the following:

- Eliminate subjective choices.
- Produce the same production process for the same parts.

- Use data that are already available within the system.
- Simplify the process of production process design.
- Allow for manual intervention when required. (Assist the engineer but do not restrict his/her innovative skills.)

These objectives require well-designed, usable systems that contain a data base of standard or existing production processes. The data base catalogs in a convenient and appropriate way for the engineer to logically combine or adapt standard processes when designing a new production process.

Approaches to Process Planning

While computer-aided process planning systems can be complex and sophisticated in functionality, a good CAPP system is not complex to use. All systems are based upon the assumption that there is an orderly, logical method of assessing production needs and that, although there may be many ways to achieve those needs, optimum methods are available using standard existing processes. Of course, innovative new processes may be associated with a new product, and the CAPP system has the flexibility to allow new methods to be introduced and perhaps become a future standard. However, even when new processes are introduced, many of the ancillary processes will be standard or adapted standards.

There are a number of ways to approach the issue. Traditional methods require the engineer to manually examine product drawings, identify similar products and similar processes, retrieve these processes, and then modify or adapt them to his/her purpose. The advantage of the traditional way is its flexibility. The disadvantage is its lack of consistency. An improvement to this traditional method is to have a "workbook approach." The workbook prescribes the process-design process and provides the engineer with detailed and well-catalogued cross-reference information for the selection of standardized processes. This approach can be very successful with simple products and small numbers of processes. However, larger organizations find that the volume and complexity of the data required for the workbook make this method impractical and difficult to keep up-to-date.

The "variant approach" to product process design uses a CAPP system to store the detailed information and, using group-technology coding of the processes, identifies similar products or similar processes and makes them available to the engineer for modification and editing. With the variant approach, the CAPP system has no inherent logic; it is merely interrogating product and process catalogs.

The "generative approach" to CAPP system usage, instead of trying to find similar products and processes, creates the production process from predefined process algorithms. In other words, the system breaks down the physical actions that must take place for the product to be manufactured and then, using logic built into the system, creates the production process required. The standardization comes into the process, not from comparison with other products, but by the consistent use of the same algorithmic approach to designing the process.

Algorithms required for a generative approach to CAPP are based upon the use of decision trees and logic tables. In effect, a series of ever-more-precise questions are asked by the system about the production-process requirements of the new product. The system steps through a logical path to determine what processes are available to fulfill the process requirements, arriving eventually at a standard solution. This approach was pioneered by a consortium of Scandinavian companies in the 1960s.

A combination of the variant approach and the generative approach is called the "axiomatic approach." The CAPP system contains standard production processes for product families. The system selects the appropriate family and then uses decision trees and tables associated with that family of product. This method is more flexible because the CAPP system can then have a wide variety of different decision trees catering to different types of products.

A further refinement of these concepts is the "constraint-based approach," which seeks not only to develop standard processes but also to take account of the constraint of the materials and the product itself. This procedure results in what can be thought of as three-dimensional decision trees using Venn diagrams, where a solution is obtained when the common aspects of material, design, and processes overlap. These systems are highly sophisticated and complex, making use of techniques of expert systems and artificial intelligence to make judgments between conflicting criteria. The use of constraint-based systems is limited at present. However, as they are developed and their use simplified, they will become powerful tools for companies wishing to design standard processes automatically and quickly.

Programming Numerically Controlled Machines

Manually controlled and operated machine tools can be very flexible if the people operating them are skilled and innovative. However, highly precise and complex parts are difficult to make manually because even a small error can consign the part to the scrap heap. One cardinal rule of production quality is consistency. Consistency means making the products precisely the same every time. The tenth unit is the same as the thousandth unit which is the same as the millionth. While this level of consistency with machine-tool processes is difficult to achieve with manual operations, the introduction of automated machine tools has opened up the possibility of near-perfect consistency leading to massively superior quality.

Initially, automated machine tools were dedicated to just one task or a small range of tasks. They were used on production lines requiring the manufacture of large quantities of identical products for mass production. Consistency was gained but flexibility was lost. In the late 1950s new kinds of automated machine tools were developed that could be programmed to do a variety of tasks. By today's standards these machines (developed jointly by the U.S. Air Force, the Aerospace Industries Association, and some academic institutions) were rudimentary, although they used a programming language to define the geometry of movement of the cutting tools, the spindle speed, and other machine controls. These machines over time became more sophisticated and programmable machine tools (or numerically controlled machines) and are now commonplace among Western manufacturers.

In most organizations using NC machines, the programs are written by hand. A skilled technician studies the drawings and production process sheets from the design department and determines the correct method of performing each manufacturing step. He or she then writes a program to instruct the machine step by step how to play its role in the manufacture of the part. In recent years two programming languages have been widely accepted for NC machines. They are Automated Programming Tool (APT) and Compact II from Schlumberger Technologies, a manufacturer of NC equipment. These languages are designed to express the programs using terminology and syntax that is familiar in style and approach to engineering practice. Instructions are entered in terms of geometry and movement that can be read from a drawing. Compact II is specifically designed to meet the needs of smaller, simpler machines and is deliberately less complex a language than APT.

Even though these languages are easier to use than generic programming languages like Fortran or C, the programming process is still largely manual. It requires technicians to study the drawings, precisely understand the details of the process, and write the APT or Compact II program to meet the need. The program, once written and tested, can be stored (digitally or on punched strips) and used over and over again. The NC machine can often be changed over very quickly because the changeover requires merely exchanging the programs. It does not require the removal, exchange, calibration, and testing of tools or dies.

In the 1970s new methods were developed to allow NC machines to be programmed in a more intuitive way. Instead of writing programs by hand, the programmer presents the required movements to the machine graphically, using a CAD drawing of the part and directing the machine's movements with a pointer or a mouse. Using these techniques, technicians no longer have to study the drawings and translate them into tool movements; they can program the machine in a more tangible way. The result is better and much faster programming, and the machines can perform more complex procedures. Such graphics-based programming techniques have radically changed the way machine tools are controlled. While they are not yet fully accepted among manufacturers (many of whom still have machines programmed in APT), eventually interactive, graphics-based programming will become the standard method.

The degree of control and flexibility of an NC machine affects the complexity of the programming. Early machines could be controlled only through two or three dimensions in the geometric movement of the tools. Four-dimensional or four-axis machines allow 3-D control of the tool plus control of the rotary movement. Five-axis control — the most sophisticated at this stage of NC machine development — is the same as four-axis control with the additional control of the angular movement of the spindle or table. Four- and five-axis machining was originally available only within the aerospace and automotive industries because the machines were prohibitively expensive and the software inordinately complicated. In recent years the software has become easier to use and less expensive, and the machines themselves have come down in price. Sophisticated multiaxis NC machines are now being used in all kinds of manufacturing companies throughout the world.

The development of graphics-based NC programming techniques and the use of CAD and CAPP systems have led to the ability to automatically create NC programs from CAD drawings. CAD drawings contain all the geometric information required to instruct an NC machine how to manufacture a part. In fact, a CAD system can often "visualize" the complex movements of a four- or five-axis machine tool better than a trained technician. These developments make it practical to design a product and program a production plant to make the item. Therefore, new realms of possibility open up for the agile company seeking to provide highly flexible and responsive customer service in a design-to-order production environment while maintaining 100 percent quality.

A recent study from Lehigh University in Pennsylvania envisioned a 21st-century automobile manufacturer providing a service whereby customers would design their own dream car using a CAD system in the showroom. The custom car would be delivered in three days and the company would guarantee quality and reliability for the vehicle's entire lifetime. While this scenario may seem futuristic, and it is, there are companies moving in this direction already. The use of sophisticated, integrated software is essential for the achievement of this level of improvement.

Concurrent Engineering

Sometimes referred to as simultaneous or parallel engineering, concurrent engineering (CE) seeks to shorten design lead times, to improve quality, and to lower costs. These noble objectives are achieved through a team approach to product development and the use of disciplined design techniques.[2]

Agile manufacturers embark upon concurrent engineering for the following reasons:

- to design products to meet customers' needs and wants
- to shorten time-to-market for new products
- to enable a fast break-even point
- to incorporate fewer changes in design late in the project
- to manufacture products more easily and cheaply
- to make high quality the top priority
- to lower the product's total cost over its life cycle
- to lower the risk of failure

Just as agile methods change the culture on the shop floor and represent a new way of life for production personnel, so concurrent engineering is a radical and total change in the way products are designed within the organization. (See Figure 10-1.) It is not a fashionable new method; it is a change of culture in the area of product development. Design projects are completed by full-time multidisciplinary task forces drawn from many areas of the organization, including people from design engineering, manufacturing engineering, marketing, purchasing, finance, and vendors of raw materials, components, and production equipment.

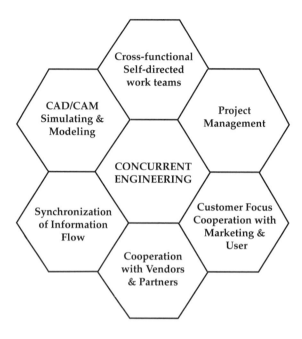

Figure 10-1. The Changing Culture of Concurrent Engineering

The combined skills, backgrounds, and perspectives of these people make for a better-designed product that meets customer needs because sales and marketing people are on the team. The product can be put immediately into production and manufactured to high quality standards because manufacturing engineers and quality personnel are on the team. The contribution of various vendors and purchasing people not only ensures that the components, raw materials, manufacturing technology, and outside processing fully meet quality requirements; it also brings the outside companies into close partnership. Financial people work with engineers and marketing people to match production costs with the price the market can bear to ensure that the company obtains adequate return over the product's lifetime. If appropriate, people from the distribution organization, field service, environmental engineering, and legal services can be part of the team and bring their own important perspective to ensure the very best product in the shortest possible time.

One objective of working together in a cross-functional team is to have various design and development tasks completed in parallel. Instead of manufacturing engineering following design engineering, these two aspects of design are completed together. This not only reduces the amount of time needed to design the product, but also provides a better design because experts from the two areas work together. Also, as illustrated in Figure 10-2, design costs decrease because it is less expensive to make changes in the early stages of design rather than later in the development cycle. Early changes mean less disruption and less wasted effort pursuing activities that are later cut out.

Despite being a relatively new approach, concurrent engineering has had some notable success within U.S. companies. Northrop, a large defense contractor, has used CE to reduce in-line modifications by 75 percent, work-in-process by 50 percent, and production cycle times by 33 percent. Digital Equipment reduced the time-to-market of new computer products from 30 months to eighteen months by using CE, and Digital is now targeting a twelve-month development cycle. Digital credits CE with product-cost reductions of 50 percent and sales increases of 100 percent, bringing the break-even point of Digital's 3100 Series VAX computer forward by at least six months. Cadillac reduced the redesign time for its Eldorado model from 80 months to 55 months. Chrysler's Viper series of sports cars achieved a total design time (from concept to showroom) of less than 40 months.

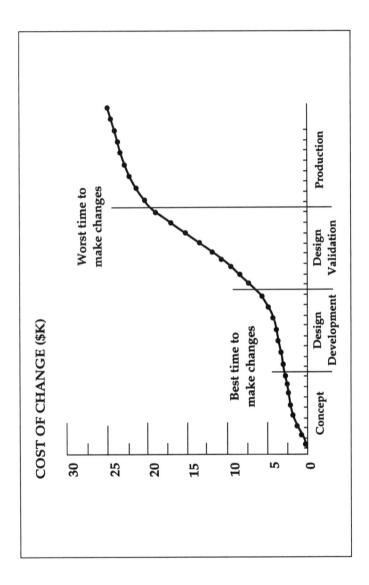

Figure 10-2. Concurrent engineering allows you to make design changes earlier in the design process.

These successes represent tangible improvement in profitability and customer satisfaction. Getting to market first with innovative new products can mark the difference between a company's success or failure.

Similar gains are being made by European companies. Rover Group, the British automobile manufacturer, used to expect six to eight years to develop a new product with a further six to twelve months for pilot production. Rover, a company in grave danger of total failure, created a joint venture with Honda and introduced CE as a part of the partnership arrangement. Rover engineers are now confident they can achieve a 48-month development cycle. Their new Discovery four-wheel-drive vehicle was designed and in production in 27 months (although it was a major enhancement rather than a totally new vehicle). Volkswagen (VW) introduced concurrent engineering in 1987 with a view to improving quality and reducing time-to-market. It also introduced massive quality improvements on the shop floor resulting in the disbanding of the quality-control department in favor of building quality into the production process. VW can now claim a time-to-market figure of 36 months, with a target to reduce it to 32 months. Other companies like Perkins Engines, General Motors's Opel subsidiary, Bosch, and Lucas Industries (a supplier of automobile electrical equipment once known derisively in England as the company that invented darkness) have all made massive improvements in quality, time-to-market, flexibility, and customer service.

Project Planning and Control

One of the most difficult aspects of concurrent engineering is the project planning associated with it. The design team comprises people with different backgrounds and skills. They are also brought together solely for the purpose of designing a certain new product or range of products. The team must be well led, well organized, and well controlled. Much of what it takes to bring a product to market quickly involves eliminating (1) the waste and delays in the design process and (2) the overlapping of tasks being performed. It also requires the evolvement of commitment in the team. This kind of project control takes skill, experience, and flexibility.

Project-planning software can be useful in helping managers bring the project under control. This software enables tasks to be broken down into smaller tasks with dates and responsibilities associated with them. Estimates of the time required to complete each task can be entered into the project-planning system. The system will then calculate when the project can be completed and will highlight bottlenecks and slack times within the project plan. The software prints time-line charts, critical path charts, and other graphical methods of displaying the project plan. In addition, this software can be useful for communication between the team members. The software will print task lists for each person and can update the task when it is completed fully or partially.

This software can be a simple graphics package that essentially draws the time-line charts, or it can be a highly complex planning, tracking, and communication tool capable of controlling the thousands of activities required to build a battleship or send someone to the moon. Project-planning software is available on PCs, workstations, minicomputers, and mainframes and is sometime incorporated into CAD systems.

While these project-planning tools are helpful, even essential for a large project, there is a danger of "paralysis by analysis." We have all known project managers who spend their time poring over the plans, entering results, updating the schedule, replanning the time line, and adjusting the critical path. The planning process is just a tool — it is not the project's objective. This danger is not a function of the software but rather in the way the software is used. Project-planning software, particularly when part of the network used for CAD and other development systems, can be a valuable communication tool and can contribute significantly to building both the team and the product.

CAD Systems and Concurrent Engineering

For most companies, CAD systems are essential for concurrent engineering. They not only speed up the design process and produce the drawings more quickly and easily than by the use of conventional drafting, they also provide a vehicle for team interaction. The diverse group of people on the team bring different needs and skills to bear on the project. The combination of these skills, plus the overlapping of design tasks, leads to CE's success.

CAD's essentially visual approach, particularly the 3-D rendering of the products under design, makes for fast and thorough communication of product specifications and characteristics. CAD systems often have modeling and simulation features specific to the kind of product being designed. These features enable the team to work together concurrently on many aspects of the design. The team should be able to simulate as many different alternatives as possible on the computer screen instead of having to wait for the production of a physical model or prototype. By the time metal is cut, every practical option can be reviewed, discussed, simulated, and evaluated. This system improves the design's quality, speeds up the design process, and reduces the number and size of changes made at the latter end of the project.

Quality Function Deployment

One of the techniques of concurrent engineering is quality function deployment (QFD). Quality function deployment is a formalized method of matching the expressed needs of the customer to the features and functions of the product.

Classic QFD uses a diagram called the "house of quality," which lists the customer's expectations of the product down the left-hand side of the chart. The planned product features are shown on the chart and matched to customer needs. Other aspects like competitive analysis, function interaction, and priorities are also shown on the chart. This sophisticated visual method keeps the designer's eyes squarely on customer needs and is a powerful CE tool.

Essentially graphical, QFD requires the drawing of house charts that match customer needs with product features. When first devised, QFD charts were drawn by hand and posted on the walls of the design area so that everyone could see and use them to focus discussion and team interaction. Now computer packages that either stand alone or are incorporated into CAD systems hold the information on a data base and then print the charts in various ways. These systems also produce other analysis reports. The advantage of these systems is that they make the drawing of the charts quick and easy. They also provide analysis that would otherwise have to be done by hand. Their disadvantage is that they can take away the team interaction in the chart's development. The objective of QFD is *not* to draw the chart; it is for the team to work together for the coalescence of customer needs and product features. It does not matter if the chart looks chaotic or has to be redrawn several times. However, many companies successfully combine both the convenience of a computer system to collate the data and draw the charts and the team interaction so vital to QFD's success.

Design for Manufacture and Assembly

A classic problem of traditional product design is that design engineers lack sufficient knowledge or regard for the problems of manufacturing the products they design. The result inevitably necessitates that substantial changes be made to the design through the production-engineering and pilot-manufacture processes. One big improvement coming from concurrent engineering is that the product's manufacturability can be brought into the design at its earliest stage. A number of aspects are incorporated into the ideas of "design for manufacture and assembly" (DFMA). These aspects include designing components and assemblies that are physically easy to make and do not require either machines or operators to become contortionists, designing components and assemblies that can only be assembled in one way (failsafe), and designing that lends itself to cellular manufacturing. The objectives of DFMA can be summarized like this:

- Use a minimum number of component parts.
- Use modular design.
- Design parts to be multifunctional.
- Minimize part variation.
- Differentiate products at the latter stages of manufacture.
- Use standard parts.
- Design parts that can be manufactured easily.
- Use clips and single component fasteners instead of separate fasteners.
- Minimize handling.
- Eliminate or simplify adjustments.
- Avoid flexible materials.

These principles are not hard and fast rules; they are guidelines used by the design team to improve quality, reduce cost, and standardize products. All components and subassemblies are examined with a view to eliminating, simplifying, or standardizing the item. Software is available that assists the engineers to review each component and to study its manufacturability. Each component is given an assessment on a scale of zero through nine according to its particular design criteria. The sum of the assessments is a gauge of the product manufacturability. The purpose is not to gauge manufacturability — the purpose is to improve it.

Failure-Modes-and-Effects Analysis

The purpose of failure-modes-and-effects analysis (FMEA) is to examine product designs to determine places in the product where failure is likely to occur and to minimize that risk. Each component or each aspect of a component is assessed for (1) its probability of failure, (2) the seriousness of the failure, and (3) the difficulty of detecting a failure. These three factors are combined to create a criticality index for the items. The team then gets to work on the components with a high criticality index to eliminate the likelihood of failure.

Software is also available that assists engineers in their assessment of the likely failure of parts and subassemblies. The software does not do the job for you; it just provides a standard approach and a method of analyzing and disseminating the information. However, it can be valuable, particularly where the products are large or complex and a substantial amount of analysis is required. The volume of data is often a more significant problem than the complexity of the data and analysis.

Analysis similar to FMEA and DFMA can be accomplished for the production process as well as for the product design itself. Some agile companies use the Taguchi method for analyzing each step of the production process, for understanding how many different parameters are involved in the process, and for reducing the likelihood of failure or quality problems by using visual charts and graphs to brainstorm solutions.

Target Costing and Value Engineering

Target costing is a technique developed by Japanese automotive companies to incorporate financial analysis into the team approach to product design.[3] Although product design in traditional companies includes estimating the cost of manufacturing the product, target costing is used as a systematic approach to understanding the drivers that create cost and then establishing cost targets for the design engineers. Target costing coupled with value engineering (VE) provides the tool for bringing product costs under control during the design phase of the product life. The tendency of traditional companies to embark on cost-reduction programs after the product is in manufacture leads to many changes in the components and processes during the first few months of production. This method is expensive, creates confusion, and often reduces quality. Target costing's intent is to go through the cost-optimization process during product design.

Target costing begins with the marketing people assessing the market price for the product under design. From this assessment the financial people calculate the "allowable product cost" by subtracting the margin the company wants from the product. The objective of the design team is to design a product that fully meets customers needs, can be manufactured to very high quality levels, and has a production cost no higher than the allowable product cost. In the early stages of design, target costs are calculated at a high level — for example, the whole car or major elements of the car, like the transmission or the motor. As the design progresses, target costs are established for lower-level items until at the final stages of design the target costs reach down to individual components, materials, and subassemblies.

Target costs are independent of the design and derived from the market price. Accountants on the team will also calculate the actual cost inherent within the design. This cost is known as the "potential cost." Invariably a gap exists between the allowable cost and the potential cost. It is this gap that must be closed through the design process if the product is to meet its profitability objectives. The accountants on the team then establish target costs for major product components, and the team gets to work reducing the potential cost through value engineering and other methods. The target cost is not always made to equal the allowable cost, because the allowable cost is the production cost of the product in the medium-term time frame. There may be some reason for allowing early production costs to be higher, in which case the target costs will reflect this compromise. Target costs are always carefully established to be both challenging and attainable so that they are motivational and not discouraging.

Value engineering is an approach to product design that seeks to reduce costs inherent within the design of a product and bring them into line with the target costs established for the product. Of course, all companies have procedures for reducing product cost through review of the components and processes that make up the cost. Value engineering is a thorough and systematic approach to the analysis of cost drivers and the reduction of those drivers throughout the design process. In the same way that QFD systematically keeps designers focused on the needs of the customers, so VE keeps the designers focused on bringing potential costs into line with target costs.

Concerned not only with the costs of the product being designed, VE takes a wider view. This wider view includes an assessment of lifetime costs. Lifetime costing is concerned with maximizing profits on a product or product line over its entire life. The lifetime view will often result in design changes that are beneficial over the product's lifetime even though they are negative in the short term. A lifetime view of costs and profitability will often take account of product costs from the customer's point of view rather than just the manufacturer's. This overview will include an understanding of the ownership cost of the product and, where applicable, the cost of final disposal.

Another important aspect of value engineering is the overall cost to the company. It may be that selecting a particular component or process will reduce the cost of the product but will not be in line with the company's broader goal of standardization and part reduction. Designing the product to be modular may not be the most cost-efficient approach for the product being designed, but it may be a broader company objective. These issues need to be included in a systematic way as design engineers and accounting people apply the principles of target costing and value engineering.

Many Japanese world-class companies perform these target-costing tasks manually on charts or by using straightforward computer spreadsheets. In many instances, it is not necessary to target cost every single part and subassembly. It is often more useful to focus on the problem areas in detail and handle the other targets at a macro level. Here, the software to support target costing can be helpful by collating large volumes of information and presenting it to the team. Also, there are cost issues that affect several individual parts or subassemblies. The software can immediately update the cost changes affecting many parts or subassemblies rather than requiring someone to change each one manually.

Target costs can also be included into the cost buildup system. Most cost-accounting software has a cost buildup feature that reads through the bills of materials and production-process routings to determine the potential cost of the product. In most companies, cost buildup systems are used to determine the standard cost of current products. For products under design, target-cost information can be held on the files holding potential costs and can be used to report the cost gap for each item and subassembly. These cost gaps can be added and summarized up to major subassemblies and to the final product.

Target costs can also be used after the design is complete and the product passes into the production phase. More helpful than standard costs for continuous improvement, target costs can have more cost elements, they can vary from one location to another, and they can change over time. Most standard-cost systems have a limited number of cost elements (such as material, labor, outside process costs, fixed overhead, and variable overhead). Target-costing systems have a larger number of cost elements, and these elements can be used for cost drivers that are unique to the product concerned. There can be more than one target cost for a part or product because the method of production may vary from one location to another (if the product is manufactured in more than one place). Similarly, the target cost may change over time as continuous-improvement projects bring the organization greater production effectiveness. These improvement goals can be incorporated into the target costs once the product is out of the design phase. Software supporting target costing can have effectivity dates built into it so that the changing targets can be stored and used.

Re-engineering Existing Products

A great deal of attention has been given to innovative approaches to product design. In reality, however, most companies have existing products that will be the bread and butter of their operations and profitability for the foreseeable future. While new products will be introduced from time to time, existing products are the ones that need to be addressed for their world-class suitability. Some industries, such as consumer medications, do not have innovative new products being introduced on a regular basis. All new work in consumer medications is taking place in the pharmaceutical companies and will be introduced initially as prescription medicines before being marketed as over-the-counter remedies. A company must become a world-class manufacturer of existing products. It cannot wait for new items — concurrently engineered, designed for manufacture and assembly — to be introduced.

Process Analysis

The starting point with many agile manufacturers is that quality and production effectiveness come from perfecting the process of manufacture. This goal of striving for perfection is equally true of the distribution process, the clerical-support processes, and all other activities within the company. All business activities are processes to be perfected. Dr. W. Edwards Deming points out that only 15 percent of quality problems in a production plant are caused by production operators misjudging or making mistakes. The remaining 85 percent (the majority of quality problems) are caused by managers relying on inadequate processes. And the managers are responsible for the processes.

Because a cardinal aspect of agile manufacturing is perfecting the process, two approaches must be addressed: (1) the quality of the process and (2) the waste within the process. From the perspective of quality, statistical process control (SPC) is one of the most powerful techniques for perfecting the manufacturing process. SPC enables operators to monitor a process's ability to consistently make products in accordance with design specifications. SPC shows whether the process is under control or out of control. Eliminating waste within the process requires a detailed understanding of what the process does and what its constraints and variables are. To gain this understanding, the process must be analyzed and "mapped."

Process mapping is a systematic method for gathering the knowledge of people within the company about a particular process. A process (for example, raising and actioning engineering-change notices) may involve personnel from many different departments. In traditional companies it is unusual for anyone to fully understand an entire process, let alone have a detailed appreciation of the constraints and variables involved. The technique of process mapping brings together everyone involved with the process (plus other personnel on the cross-functional team) for the purpose of fully understanding the process and eliminating waste.

The start of a process-mapping exercise is to define which process is to be studied. This may sound obvious, but keeping a mapping project focused on just one aspect of the business can be difficult. It is important to define where the process starts and stops before trying to understand it. The next step is "storyboarding." Storyboarding is a brainstorming session among the team members, using a visual wall chart to bring out all elements of the process under consideration and to fully recognize how these elements interact with each other. For even a straightforward process, this storyboarding activity can be complex and lengthy. The interim result of storyboarding is to draw a process map of the process. A process map is a flow chart, using standard symbols and nomenclature, showing and explaining all the steps involved in the process. The explanation of the steps will include information like the following:

- How is it done?
- Who does it?

- When do they do it?
- Why do they do it?
- Is it a decision point?
- Is it a delay?

The process map fully describes the current process (as shown in Figure 10-3). However, the purpose of the exercise is not to draw a map of the process, but to bring together the entire body of knowledge about the process to create a vehicle of communication that is visual, and to use the result to bring about process improvement. The cross-functional team works together to review and critique each step of the process. Does this have to be done at all? Can it be done better? Can a delay be eliminated? Can the process be synchronized to eliminate bottlenecks? Can steps be overlapped or paralleled to reduce the process's cycle time? The process maps do not have to be one standard nomenclature. Maps can be used to show not only the process but also the costs, the time taken by each step, the departments responsible for the task, or other pertinent information. Every and all possibility is examined systematically, and new approaches to completing the process tasks are suggested and evaluated.

The team then produces a chart of the new and improved process. The new process can be evaluated through a "conference-room pilot" — a simulation of the entire process with team members carefully working through each step to ensure that the new process really does cover all the required activities. By using the new process map as an education tool, the new process can then be put into action with considerable elimination of waste.

In companies using activity-based costing, management accountants already will have considerable knowledge of the processes, in particular the activities that drive a product's costs. This factor makes management-accounting

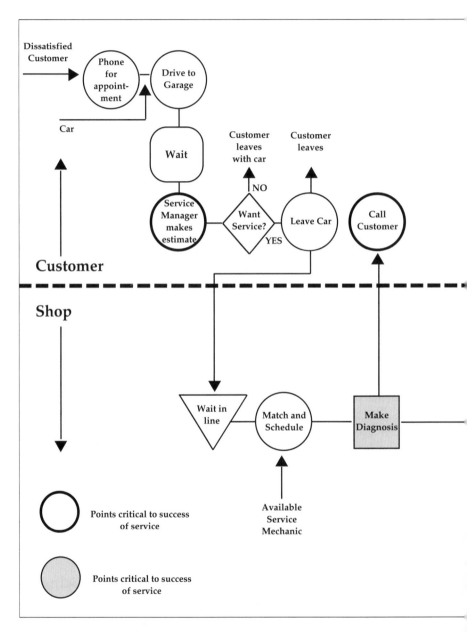

Figure 10-3. An Example of Processing Mapping: Car Servicing

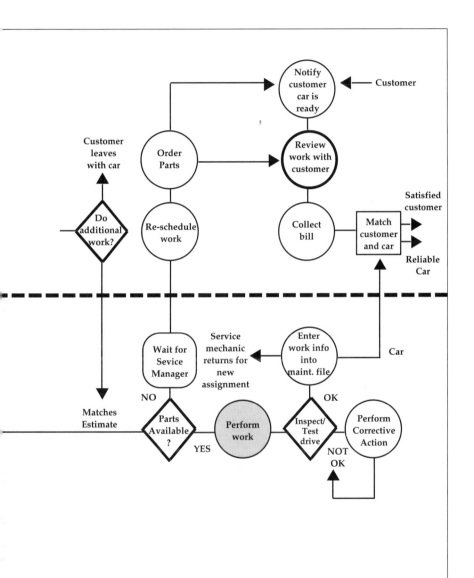

An Example of Process Mapping: Car Servicing

people ideal members of process-improvement teams. In addition to their detailed knowledge of precise activities, the accountants' analytical skills are valuable for understanding and improving the process. The use of ABC information as a resource for process improvement is a powerful new tool for the agile manufacturer.

Process mapping is an important method for assisting cross-functional teams in the quest for perfect production (or other) processes. It is essentially a graphics, open-communication method. Software can help in drawing the charts and in collating the tasks and activities, particularly when tasks are repeated across many different processes. Having a data base of the activities (coded according to who does them, when, and where) can also be useful. This data base can be analyzed and the improvement activities expedited. However, it is important not to let the software create barriers between team members. The graphical nature of the exercise builds teamwork that can be hampered if someone is poring over a keyboard while the others are excluded. Some companies have stumbled due to an overconcern with producing attractive charts instead of concentrating on the improvement process. The purpose of the charts is to help bring process improvement. The purpose of the software is to remove unnecessary work and to assist with the process-improvement tasks.

Process-mapping software can be helpful providing it is used to collate the information and draw the final charts. However, it must not replace the team storyboarding and brainstorming activities. Often the information contained within the computer system used for activity-based-costing analysis can be a foundation upon which process-mapping and waste-elimination projects are built. Although ABC looks at the company's process from the point of view of cost drivers, it has collected much of the data required for process improvement.

Value-added Analysis

Value-added analysis is an extension of process mapping designed to identify those steps in a process that add value to the products being manufactured and those steps that are wasteful. A value-added step is defined as an activity that adds value to the product from the customer point of view. Non-value-added steps are everything else the company does. A typical U.S. company doing a detailed value-added analysis of its production operations finds that of the

---

Value-added and Non-Value-added Activities

Value-added Activity
an activity that transforms or shapes a product or service toward that which satisfies the customer's real and perceived wants and needs

Non-Value-added Activities
Strategic
an activity that indirectly transforms or shapes a product or service and is strategically important to the long-term growth of the business

Support
an activity that does not add value as perceived by the customer but provides a service that maintains the natural process of the operations or is driven by actions outside of the organization's control

Controllable (Waste)
an activity within the control of the organization that consumes time, resources, or space without contributing to the transformation or shaping of the product or service

time spent in a production plant, 5 percent is value-added and 95 percent is wasteful. It is the elimination of the wasteful activities that is so crucial to agile manufacturing. Traditional companies have often spent a great deal of effort studying production activities and improving their efficiency despite the fact that they are only 5 percent of the total time spent within the plant. The real savings and improvements can be made in the 95 percent non-value-added activities.

Value-added and Non-Value-added Activities

Some non-value-added steps are necessary. Typically, these activities are support activities like production planning and control, accounting processes, and the like. While in the longer term, many of these activities will be considered wasteful, in the short term they are considered non-value-added but essential. Other activities may not add immediate value to the product and services — but they will in the long run. These strategic activities include such things as product research and development or education and training of the work force.

The prime purpose of value-added analysis is to identify the areas of the company's activities that are waste and to assist the process of eliminating waste from the organization. This technique is used by the cross-functional teams assigned the task of eliminating waste. The teams will analyze processes using process-mapping techniques and assign a value-added category to each task. The prime candidates for elimination will be those tasks assigned a category of "waste," which will be examined, evaluated, and systematically eliminated from the process. Some activities, even those considered entirely wasteful, can be difficult to eliminate. Often the process-improvement teams make several passes at a process before eliminating a substantial amount of non-value-added activities. The quest for waste elimination is not a quick fix — it is a long-term commitment to excellence and continuous improvement. The tools and techniques of process mapping, value-added analysis, and process improvement are used continuously throughout agile manufacturing organizations.

Computer Systems and Value-added Analysis

The production routings contained within the production planning and control system or the engineering system show the precise steps required to manufacture each product. The steps involved will include both value-added activities (like assembly or fabrication) and non-value-added steps (like inspection, material movement, and wait time). As a value-added code is assigned to each step within a production routing, a report can be written calculating, summarizing, and displaying the value-added and non-value-added activities. Often within engineering systems, individual production activities are expressed as operation codes that relate to an operation-code table or file containing all valid operations. In this case, a value-added code is placed in the operations-code file, allowing all the activities contained within the production routings to be easily classified for value-added and non-value-added analysis. Of course, this procedure will only show activities included within the production process and does not deal with any ancillary and support activities. Nevertheless, the information is readily available and pertinent to improving production.

Once again the information gathered and used in an activity-based-costing exercise will be most helpful for value-added and non-value-added analysis. In fact, many software packages available to assist with ABC have the ability to assign a value-added code to an activity and are able to print various value-added reports and charts. The analysis required to complete a thorough ABC study requires that the accountants involved in the study become knowledgeable of the production processes and have a detailed understanding of which activities are value-added and which are not. This is yet another instance where management accountants can be effective team members for value-added analysis and process improvement.

Variety Effectiveness Program

There is positive variety and negative variety within a range of products. Positive variety incorporates a wider range of choices of finished products being offered, a feature that customers value. Negative variety occurs within finished products where the choice range is too wide and which customers find complex and unhelpful. A second form of negative variety occurs when the needs for finished-product variety create uncalled-for complexity in the raw materials, components, and subassemblies throughout the production process.

An example of negative finished-product variety is taken from a Canadian company that manufactures drilling bits for oil rigs and other deep drilling facilities. This company once had approximately fifteen product families of drilling bits, each containing several different kinds of bits. These families each had different engineering characteristics. A product-rationalization program revealed that customers were bewildered by the variety of different bits available and were uncertain when to use each kind. The company redesigned its products, reducing its product range to only three product families. Each of these families was then re-engineered to fill a broader range of tasks. The customers were happy because they only had to stock three different kinds of drilling bits instead of fifteen. In turn, the company gained substantial savings by making, stocking, and marketing far fewer products.

In an example of negative variety within the internal operations of an organization, an electronics company required over 1,500 different kinds of resistors. Of these, there were 120 different kinds, sizes, and tolerances of 1,000-ohm resistors, which were used across many different products the company manufactured. An analysis of the usage of these resistors showed that the 120 variations could be reduced to only sixteen by re-engineering the products with a view to increasing the commonality of components. The company eventually went from 1,500 different kinds of resistors to less than 200 actively used for manufacturing. The savings associated with this reduction of components and increased level of commonality were enormous. Many of these savings were intangible and difficult to quantify, as they included the reduction of confusion and complexity within the procurement, inventory, and production-control processes. Many products had slightly increased material costs because the item now contained a more expensive component. However, this cost was vastly outweighed by the broader savings that accrued from the increase in commonality and reduction of variety within the process.

The same principles apply within production processes. If the number of different processes can be reduced and more commonality of process introduced within an organization's product range, then the company will see enormous tangible and intangible costs. In addition, increased process commonality makes the company more flexible as customer product-mix and product-volume needs change.

Computer Systems and Variety Effectiveness

The problem with conducting a thorough variety-effectiveness analysis is the volume of information that must be processed. Computer systems can help with selecting which product ranges to look at first, with assessing the complexity of the product designs and the potential for simplification, and with the decision-making process. No hard and fast rules govern the correct balance between variety and simplicity. It is a subtle engineering and marketing judgment. Well-designed software can be used to analyze data, present alternative approaches, and provide clear information (often graphically presented) so that engineers and marketing personnel can make the optimum decisions. The software can also be used to analyze new product designs to prevent negative variety in the future.

Summary

An agile manufacturer approaches product design and development differently from traditional companies. The objectives of product design are to quickly develop and bring to market innovative new products that exceed customer expectations and desires, have high quality and low cost, and can be manufactured easily. These objectives are achieved through the use of cross-functional design teams employing the techniques of concurrent engineering in which time-to-market is significantly reduced by the overlapping and more time-efficient development activities. The use of a cross-functional design team also enables products to be designed for quality and ease of manufacture.

Computer-aided design and computer-aided process planning systems are essential software systems for the concurrent-engineering approach to product development. CAD and CAPP not only save time and cost by automating many traditional design-engineering tasks like drafting, but also provide computer-modeling techniques that allow the team to work together to design a much better product. CAD and CAPP support commonality of components and processes by maintaining data bases of currently used components and processes that can be incorporated rationally into new products. This function reduces complexity, reduces the learning curve, reduces costs, and enhances production flexibility.

Integrating CAD and CAPP systems with the shop-floor machines is the big step many companies wish to make so that computer-designed products can be manufactured immediately by passing information electronically from design to production equipment. While such integration is possible and some companies have been successful with CAD/CAM, the result is still complex and difficult to achieve because standard methods of communicating with and programming the shop-floor machines have yet to be perfected. Over the next few years this level of integration will become commonplace.

Agile manufacturers have developed new techniques that put the new design principles into practice. These techniques include quality function deployment to match customer needs with product features, design-for-manufacturability analyses, and target costing and value engineering to assist with designing the products to achieve the company's profit objectives and market prices. Existing products must also be analyzed for process improvement, and cross-functional process-improvement teams can use techniques like process analysis and value-added analysis to determine potential process improvements and to put these improvements into action. Variety-effectiveness (or variety-reduction) programs are used to increase commonality of components and processes and to reduce the production complexity within the product range. Each of these techniques can be supported by well-designed and appropriate software that can enhance, speed up, and simplify these various improvement initiatives.

CHAPTER ELEVEN

Hardware, CIM, and Data Collection

*T*his book is concerned primarily with the management of manufacturing companies making the transition to world-class operations and specifically with how well-designed software can be helpful in that process. It does not attempt to deal with any of the associated computer technologies. Computer technology, both hardware and software, is changing so fast that to predict where the industry is going over the next few years is risky. However, there are emerging trends, and this chapter deals with how they relate to the use of computer systems within an agile manufacturing organization.

Our discussion begins by considering the different types of hardware commonly used by manufacturing companies, moves to discuss changes taking place in operating systems, and finally deals with some of the more important trends in software technologies. These software technology trends include the use of graphical user interfaces, relational data bases, client-server and network technologies, multi-platform functionality, and the use of open systems through development standards.

Types of Hardware

Computers once were monstrous mainframe creatures housed in huge rooms and surrounded by whirling tapes and flashing lights. The mainframe computer, and IBM its progenitor, dominated the computer business from its inception until the mid-1980s. Since then several revolutions have occurred in hardware technology, operating systems, and computing philosophies.

The first big change was the move toward minicomputers, spearheaded by Digital Equipment, which developed machines that were smaller, more robust, and easier to use than mainframes. These machines, initially used mostly in scientific and engineering operations, by the late 1980s were common within manufacturing companies. As time went by, minicomputers became more powerful and the distinction between minis and mainframes became more philosophical than real. Hewlett-Packard, Honeywell, IBM, and Digital Equipment all penetrated the manufacturing marketplace with relatively inexpensive, powerful machines.

The advantage of minicomputers did not just relate to hardware. Operating systems and system software developed for these machines were easier to use and more flexible than those of mainframes. The VMS operating system and its ancillary support software provided Digital Equipment's VAX range of minicomputers with a sophisticated, powerful, and easy-to-use operating environment. A large VAX is comparable in power and performance to a traditional mainframe machine. The IBM Series 36 machines were engineered to provide an operating system that could be used by a nonspecialist; they made use of simple menus and prompts to make the machine user friendly. IBM brought across the same approach and many of the features into the more powerful AS400 range of minicomputers that were introduced in 1988 and have been extraordinarily successful in the marketplace. When the AS400 (or Silverlake, as it was first known) was first introduced, demand far outreached IBM's production capability. It took many months before the company was able to achieve production levels that met customer demand.

The next revolution — the introduction of the personal computer (PC) — was initiated by innovative new companies like Apple and Commodore (and Sinclair in the United Kingdom) in the early 1980s. The market for personal computers became more standardized in the late 1980s as the IBM PC became widely accepted. The technology for the IBM PC and the DOS Operating System from Microsoft soon became the de facto standard, and all computer manufacturers began to manufacture PCs — except Apple, which attempted to grab the market with its Macintosh series of elegant and easy-to-use personal computers. There are now thousands of kinds of PCs used throughout the world and a bewildering array of DOS software. Despite the significant shortcomings of the PC architecture and the DOS operating system, the IBM PC now dominates the personal-computer marketplace. The PC is becoming widely accepted for networks and client-server arrangements that take the PC out of the "personal" world and into the realm of powerful business computing.

In parallel with the emergence of the PC as the standard approach to personal computing came the widespread use of workstations. There is no good definition of "workstation," and the word is often used interchangeably for a PC linked into a network. However, in this context a workstation is a powerful desktop computer designed for use over a network (or stand-alone) for tasks requiring considerable local computing power. A pioneer in this area was Sun Systems, whose workstations became widely used for such tasks as engineering design, drafting, architectural layouts, graphic design, and other applications requiring powerful desktop machines. Many workstations are manufactured and configured for specific tasks and are supplied by the manufacturer as an entire system comprising a network of workstations, together with all supporting software and peripheral equipment like printers and plotters.

In recent years, all major computer manufacturers have added a series of workstations to their product range, including Hewlett-Packard, Digital, and IBM with its RS6000 machines. Most of these workstation machines use a standard version of the Unix operating system developed by AT&T in conjunction with some academic institutions, and many of them use the new RISC architecture. Reduced Instruction Set Computing (RISC) is a new style of integrated-circuit chip that is very powerful and very fast. Most manufacturing companies use workstations in their operations at some point, often within the design-engineering department or the drawing office.

Operating Systems

An operating system is the software the computer manufacturer builds into the hardware to control the way the computer runs. It controls how information is accessed and displayed, how the calculations are performed, how computer's memory is used, how monitors and printers are controlled, and how jobs are ordered and run within the machine. The operating system on a particular machine usually also contains a number of utilities and programs to help the computer operator control the machine efficiently. These programs will include utilities to do the "housekeeping" on the disk storage; various utilities to print the output on different kinds of printers, plotters, or other output devices; communications software that allows several computers to "talk" to each other; utilities to optimize the use of computer memory; programs to back up data onto long-term storage media; and programs to allocate the amount of computer power made available to each user. In addition, many computer companies include software to provide tools like electronic mail.

Until recently, computer manufacturers have tailored their machines' operating systems deliberately making them very different from one another in order to differentiate their products. Some machines are known to have operating systems and utilities that are very powerful, user friendly, or designed for one particular kind of use. Each company's operating-systems technology was an important — and closely guarded — part of the hardware product. One of Digital Equipment's innovations was a single operating system across its entire line of VAX machines from the smallest to the biggest. On the other hand, IBM has separate operating systems for each series. The mainframe operating system differs from the System 36's, which differs from the System 38's, which differs from the AS400's, which differs from the workstations or PC hardware.

The drawback of this variety for the user is that application software must be written to run on a single kind of machine and is not easily transferable from one machine to the next. At the same time, the advantage of different operating systems is that they have been designed and engineered to get the very best performance from the machine on which they are used. Different machines have different features and (often) different market needs; the operating systems are designed to provide for those differences.

However, today there is a big change in approach. The market has begun to require commonality of operating systems across different kinds of machines. The rapid increase in computing power has diminished the need for machine-specific operating systems. Also the customers' desire for more flexibility in choice of hardware has led to the computer companies coming together to agree on standard approaches to operating-systems technology. Computer companies have been very reluctant to comply with this development because it removes an element of differentiation between their products, and some companies have been severely affected by these rapid changes in customer needs.

PICK and Unix

Over the years many attempts have been made to create standard operating systems for computers. Many have been made by academic organizations, and most of them have failed to become accepted by a broad range of users. The PICK operating system, very successful in the early days, was one of the first operating systems to become widely used across many hardware platforms, and PICK users were often enthusiastic about its power, usability, and performance. Although of late PICK has lost its appeal and dropped out of contention as a long-term multi-platform operating system, many organizations still use PICK and the wide range of available software associated with it. Companies that sell PICK-based software are scrambling to adapt their software for other operating environments. There is even a small industry of utility software for converting PICK-based systems to Unix and other operating systems.

Unix was originally developed by two brilliant computer engineers, Dennis Ritchie and Ken Thompson, at Bell Laboratories in the late 1960s. Unix is written in the programming language C, which is available on almost every computer and is therefore portable across many machines.[1] The original developers had no intention of creating an international standard for operating systems; they were merely attempting to solve some immediate technical problems encountered in their research. Unix was made available to several universities engaged in computer-science research. The operating system was enhanced and developed well beyond the scope of the original project, emerging in the mid-1980s as a full-blown multi-platform operating system.

Every computer manufacturer provides Unix as an operating system, but unfortunately there is no one standard Unix. Two standards bodies, Open Systems Foundation and Unix International, vying for acceptance of their definition of Unix have agreed to cooperate although a single standard has yet to emerge. In addition, the major computer manufacturers have adapted Unix to their own needs, deliberately adding their own variations in an attempt to differentiate their products. Sun Microsystems developed its own version of Unix for the highly successful range of workstations that made Sun a major force in the industry. IBM has AIX, Digital Equipment has Ultrix, and other manufacturers have their own adaptations. Nonetheless, Unix has gained considerable weight in the marketplace because of its ability to run software on a wide range of machines — small ones, big ones, machines from a variety of manufacturers, and networks of various kinds.

DOS and Windows

When personal computers were first available, each manufacturer either developed its own operating system or used a standard operating system called CPM. With the introduction of the IBM PC, DOS (an acronym for "disk operating system," a standard operating system from Microsoft) soon became the operating system used on personal computers. Early versions of DOS were rudimentary and considerable criticism was aimed at both IBM and Microsoft. However, over the last ten years there have been several revisions of DOS and the operating system is now robust and powerful. The recently available DOS Version 6 considerably increases the power and flexibility of DOS and adds new features (like disk data compression) that were previously available only by buying additional utility software enhancements.

In the mid-1980s it became clear that the operating system developed by Apple for its Macintosh series of personal computers was better than DOS because it was easier to use and more readily acceptable to the customer. These advantages came from the use of graphical standard screens and the use of a "mouse" pointing device. Microsoft countered this challenge by creating the Windows software. Windows is additional software that runs on the DOS operating system and provides graphical screens similar to those of the Macintosh. In fact, legal proceedings have arisen between Apple and Microsoft in an attempt to determine if any of Apple's patents were violated.

In addition to providing graphical screens, the Windows software was also designed to overcome a significant shortcoming of the DOS operating system — memory limitation. The original DOS operating system was designed for use on personal computers having a maximum of 640 kilobytes of internal memory. These days PCs commonly have 16, 32, 64, or more megabytes of memory, yet DOS does not have an acceptable method of accessing and using this memory. Windows therefore contains many features of an operating system, including additional memory management. It also uses DOS only as a platform on which the Windows operating system can run.

This untidy compromise with its resultant loss of efficiency and integrity has led to two new approaches. First, IBM introduced an entirely new operating system called OS/2, which provides the full graphics capabilities required by the customers and replaces DOS as the standard operating system on a PC. OS/2 also contains a number of powerful features that take the operating beyond the realm of personal computing and into more sophisticated computing technologies. The second approach is the introduction of Windows NT. (NT stands for "new technology" and, despite being called Windows, is a new operating system designed from scratch by Microsoft and Digital Equipment.) Like OS/2, Windows NT replaces DOS and overcomes the shortcomings of that operating system, has a graphical approach that looks exactly like the original Windows, and provides features well beyond the needs of personal computing. Microsoft's intention is to design an operating system that will become the standard operating system for all computers into the 1990s and beyond.

Future Standard Operating Systems

At the time of this writing, where the long-term future lies with standard operating systems is unclear.[2] Unix has gained considerable popularity, but the lack of a standard version of Unix is confusing and troubling to the market, and Unix cannot be considered a powerful operating system. Unix also lacks an inherent graphical approach because it was designed before the need for such an interface was perceived. In fact, the majority of software designed for the Unix operating system is not graphics oriented.

The role of Windows NT could be crucial. If Microsoft is successful at designing an operating system that is suitable for all computers in the near future and if Windows NT gains acceptance in the marketplace, Windows NT could become *the* operating system upon which 1990s computing rests. Like all industries, it takes some time for standards to be developed, and these standards are usually developed by market needs and market acceptance rather than by engineers and technologists determining the "best" approach. What is clear, for sure, is that a standard operating system is required that will provide a graphical user interface, the ability to run software on a wide range of machines, an easy and reliable networking and client-server capability, and the ability to integrate data bases and software functionality from different software companies.

Graphical User Interface

An important trend in all computer systems is the use of a graphical user interface (GUI, pronounced "goo-ee"). Traditional computer systems have presented their information in the form of letters and numbers on screens or printouts, in which the software requires the user to type information into the system and to read information printed out from the system. A graphical user interface is easier to use because it presents the information graphically. Even though there is still a great deal of typing in data and reading from the screen, the user does not need to learn cryptic codes and instructions. The screens look and feel easier to use and the computer performs tasks by the user pointing to a message on the screen, with a mouse pointing device, and clicking a button.

This graphical approach was engineered originally by Xerox research in Palo Alto, California, in the late 1970s. It was popularized by Apple when the Macintosh series of machines was introduced. With a GUI built into the operating system, the Macintosh soon became known as the "friendliest" and easiest to use of personal computers. Workstations were designed to be graphical in their approach, and soon other companies were building graphical interfaces into their machines. A number of standard approaches were adopted by the various standards institutes, including the Motif standard for graphical interfaces.

Early attempts to introduce a standard GUI for the IBM PC failed because of the limited power of the hardware. However, the introduction of the Windows 3.0 upgrade in 1991 (and later Windows 3.1) heralded the first real GUI for the PC. Windows has been quickly adopted as the standard GUI, and many software companies have adapted their products to run under a Windows environment. Microsoft, the manufacturer of the Windows software, has also introduced a version of Windows that specifically addresses the needs of networked work groups. A host of development languages and tools are available that honor the Windows environment.

One key aspect of Windows or Macintosh software is that they all look and feel the same. Not only is the graphical approach intrinsically easier to use, but a standard style of screen design has been adopted by most software companies so that different software products from different companies all have a similar look and feel. This standardization makes it much easier to learn new software packages and removes much of the fear and complexity associated with a computer system. Graphical user interfaces are here to stay, and gradually over the next few years it is likely that all software will be written using a graphical approach. While there is still some debate on which standard will be adopted, at present it seems that Windows will become the accepted standard.

It is not essential that software used by companies moving into an agile environment have a GUI. However, it has been demonstrated in practice and by academic studies that a GUI is easier for people to use and provides a more reliable method of screen handling.[3] Providing graphical interfaces to software should not be a high priority for companies moving into agile manufacturing except where the addition of a GUI significantly simplifies the transaction processing. This is because changing systems to include a GUI is a complex and time-consuming task. While it is of great importance that the software used provide appropriate functionality, any new software introduced should incorporate a graphical user interface.

Relational Data Bases

Older systems use a data-storage technique known as "hierarchical data bases" while newer applications employ "relational data bases." Although differences between the two technologies are subtle and quite complex, the use of relational data bases offers significant advantages. A relational data base holds the data in tables represented by columns and rows. The rows correspond to traditional data-base records and the columns represent fields within the record. The relationships between two different tables are also defined within the data base. For example, if a stock record on the stock-records table contains a part number and the part number is defined on the part master table, the relational data base will know the relationship between those two records and will automatically pull together the information from both tables (and others) to produce a report or an inquiry. A classic hierarchical data base requires the software to contain information about the relationships between the records and accesses data by reading files according to their "keys." These keys are pieces of data that uniquely identify a particular record within a file. For example, the key to a part master file may be the part number or a combination of the part number and the company division using the part. The use of a relational data base simplifies the design and programming of software because the relationships between data are maintained automatically within the data base and do not require being programmed into the system.

The better relational data bases also contain useful features to further simplify programming and system-development activities. Validation rules concerning a particular piece of information can be set up in the data base so that application software does not need to include these validations when updating a record. For example, creating a part number in the part master table may require the entry of a "product class," and this product class must exist on the product-class table. The data base can check this table instead of the software having to include a procedure for this check. Similarly, entry of one piece of information may require another piece to be entered simultaneously. This double-entry procedure can be incorporated within the data base, thus simplifying the application software. A good relational data base will ensure that information is complete, is not repeated, is valid, and has the correct characteristics for the item of data.

Relational Data Bases and 4GL

Because relational data bases make information more accessible, most agile manufacturers provide users with a fourth-generation-language (4GL) report and query tool. These tools allow users to create their own reports and inquiries from information contained within the system without the help of the MIS department or other professional programmers. 4GLs can do this alone because the user is led through the process of programming the report or inquiry in a very structured and simplified way — with the 4GL tool doing complex and tedious programming tasks like formatting, defining the sort procedures, and validating data. However, a 4GL cannot automatically select information the user needs to see. The user must tell the system what he/she wants and the 4GL will find the information and present it.

When a report or inquiry is written with a 4GL using a hierarchical data base, users must have some knowledge of the data-base structure and particularly the relationships between various files within the application before they can use the 4GL tools. When a relational data base is used, understanding the structure and relationships is not so important because the relationships between the information is understood by the data base itself. From the user's perspective, the 4GL merely offers all the information available within the data base and the user chooses what to display and what not to display. The data base takes care of where the information is kept and how one piece relates to another. These features of a relational data base make the system simpler and more accessible and the application more effective. All prominent relational data bases on the market today provide 4GL tools as well as a full suite of application-development tools intended for professional programmers and system designers.

Relational Data Bases and Client/Server

The advent of client/server computing and networks requires that many people have access to the information contained within the system. A traditional system would provide this access by having the data base reside on one machine and having all users access that machine to obtain information they need. In a networking environment it is often important for the data base to be spread out across many machines. A simple example is a company with two warehouses, each containing the same products and each controlled by an inventory-control system. Although having the files relating to each warehouse reside on that warehouse's computer is the most efficient system, the other warehouse must also be able to access information from the first warehouse when a shortage occurs. Similarly, the two warehouses may have common customers who buy from both locations. Having two different customer records for these common customers just because they are served by two warehouses is inefficient; the invoicing and credit checking should consider both warehouses together.

An application using a traditional hierarchical data base accommodates this need by programming the information into the system. Unfortunately, as the application programs become more complex, they must be tailored specifically to the company's current configuration. If a third warehouse is added, the programs will need to be changed to include file calls from the third location. However, a powerful relational data base can handle this situation within itself. Tables can be spread out across the several locations, providing the efficiency of not only having the data reside locally but also allowing access from and to the other locations automatically through the data base. The application software does not need to know where the information resides; the data base takes care of that issue.

Relational Data Bases and Application Software

Several of the attributes just mentioned show that a relational data base allows much of the file-handling complexity to be moved out of the application programs and into the data base. In other words, the application software programs are simpler, easier to maintain, and do not need to know what kind of data base they are running on. An added advantage is that application software can then be written to run on several different relational data bases. The user organization is not required to buy a specific relational-data-base product to run the software, because the software already supports multiple data bases. Today all major data bases use a data-base accessing method called "structured query language" (SQL). Programs that use SQL data-base calls can be written without the programmer needing to know which data base will be used to contain the information required by the system.

Unfortunately, there is not a single SQL. As with Unix, there are several SQL standards, and they all differ. There is an ANSI-standard SQL that all the data bases adhere to. However, it has a very limited set of SQL features, and each data base contains many additional SQL features that are required for effective and efficient use of the relational data base. But these necessary extensions to the standards mean that there is no single approach to developing application-software relational data bases. In fact, application developers need to tailor programs to meet the needs of each different data base. This problem will be resolved over time as real standards are developed for data-base technology.

Problems with Relational Data Bases

There is one significant drawback to relational data bases — they are slow. A hierarchical data base used within a well-designed program can access data quickly, perform the tasks, and update the files. However, a relational data base contains a great deal of internal logic and programming. When the application software calls the data base for information, it triggers off a series of searches, validations, and other tasks required for the data base to perform the desired action. All this takes time and computing power within the machine. The good news is that computers are becoming more powerful, faster, and cheaper, making it possible to overcome the slowness of the relational data base. The second piece of good news is that developers of relational data bases have become skilled at speeding up their functionality. Every new release of a data base contains changes that speed up the operation of the software. Some data-base suppliers now claim that their data bases are as efficient as traditional hierarchical data bases (although these statements are difficult to define and verify). A data base's speed and efficiency is dependent upon what tasks the application software requires. For example, running a full MRP regeneration requires large amounts of information to be accessed, consolidated, calculated, and placed back into new files. Typically, there are thousands of occasions throughout the process when the data base is read and written back to. This kind of task will severely test the speed and efficiency of a relational data base.

An agile company should not spend the time, money, and effort required to implement a relational data base in its existing manufacturing systems. However, when making a decision to purchase or write new application software, a relational data base should be included as part of the requirements.

Client Server and Networks

Until recently, all computer systems were run on a single central machine with the users entering data or extracting data through terminals, printers, and other output devices. The terminals (and other devices) did not have any computing power themselves; they were merely used for entering and receiving information. Recent developments in computer technology enable the computing power to be spread across several machines throughout an organization.

The idea of networks is that users, instead of having only terminals, have on their desks computers that are linked with other computers within the organization. When a user needs to perform a task, his/her computer accesses the data (which may be resident on another computer) and does the transaction. An advantage of having the computer power spread among users is that each user has the amount of speed and power required for each machine. One person's work is not slowed down because another user is running a large process through the central machine. Another advantage is that users can perform other, non-networked tasks on their personal computers. Many tasks (word processing, scheduling, and others) do not require linkage with other users; these tasks can be performed on the same computers. A network can be increased or decreased in size quite easily. If the company grows and more computing power is required and more people added to the group, the network can be expanded by adding new PCs or workstations. This kind of expansion is easier, cheaper, and more flexible than upgrading to a more powerful mini or mainframe machine.

An extension of the networking idea is the development of a client/server network. A client/server network provides the best of both worlds because each user has a computer that processes information (the client machines) while a central computer (the server) provides the file-handling facilities and processes other tasks that require a larger, more powerful machine. Tasks are split between the desktop client and the central server. The server may be another personal computer dedicated to the tasks of a server, a minicomputer, or a mainframe. Similarly, the client machines (typically PCs or workstations) can be configured according to the needs of individual users.

While the use of networks and client/server technologies provide considerable flexibility, they also add complexity. A minicomputer connected to a host of terminals is a simple configuration that requires little maintenance and operational support. A local-area network (a network within a small working group) or a wide-area network (a network supporting distant departments of the organization) require a considerable amount of care and feeding. Another disadvantage is that client/server or networking require all users to have a PC on their desks, and PCs are more expensive than traditional terminals. This problem is being offset by the falling price of PCs and the need for users to have PCs for other reasons. In addition, a PC, workstation, or intelligent terminal is required for most graphical user interfaces, making the advantages of PCs over terminals generally worth the additional expense.

Between the two approaches lies a trade-off. However, software developers are all moving toward a client/server approach to computing because the market requires it. Software designed to operate in a client/server environment can also be used on a single machine with terminals; the user organization must choose the appropriate technology.

Multi-Platform Software

There is a growing requirement for software to be designed to run on many different kinds of computers. The jargon for this is "multi-platform." Ideally, the same software should be able to be run on a PC, a workstation, a minicomputer, or a mainframe. A network running the same software, in fact, may need to serve several different kinds of computers manufactured by different hardware companies. Until recently, software has had to be designed and written for one type of computer. In instances where software companies have offered the same product for more than one machine, they have manufactured separate software versions. For example, while the ManMan manufacturing system from ASK runs on either a Digital VAX or a Hewlett-Packard minicomputer, there are really two different versions of the software, each tailored to run on one machine. The advantage of having software that runs on multiple machines is that the user is not limited to one kind of machine and therefore has greater flexibility. If the business expands or contracts, the computer system can be adjusted accordingly. If the company has many divisions and locations, some of which are large and sophisticated and others small and straightforward, then the company can choose a single set of software to accommodate every location (providing the software has sufficient flexible functionality).

This issue can be approached in a number of ways, and all software companies are striving to bring their products into a multi-platform mode of operation. The first approach is to use a standard operating system that will run on many machines, its software transferable from one machine to the next. This approach is taken by the Minx manufacturing planning and control software, which runs on all the popular Unix platforms. ROI manufacturing software is also using Unix.

Another approach is to develop software in a programming language or environment (usually a fourth-generation language) that itself is available on many different platforms. Application software is independent of the hardware because it runs in the 4GL environment tailored to the specific machine the software is using. Xerox Computer Services uses this method with their CHESS product, which runs on almost any computer from a PC to a workstation to a mainframe. The 4GL technology employed is the PRO-IV from McDonnell Douglas Information Systems. PRO-IV provides the development tools, operating environment, and utilities to develop the system and run it on almost any hardware. The PRO-IV utilities also access a number of the most popular relational data bases so that CHESS can be independent of data bases as well as independent of hardware platforms. Similarly, the manufacturing and distribution data base and 4GL system from QAD, Incorporated, uses the Progress data base and 4GL designed to develop and run its application software on a wide range of platforms. Progress is primarily a data-base system that incorporates system-development tools. Recent releases of Progress also include access to other relational data bases so that Progress-based systems can use data stored on other data bases.

A third approach is to redesign and program application software using only industrywide standards. Standards are being developed for GUI technology, data-base design and access, client/server and network protocol, and other aspects of the complex work of software development. If a software developer designed and made software that strictly adhered to all of these standards, the product would run effectively on all major platforms from PCs to mainframes, because all hardware manufacturers are attempting to bring their products into line with industrywide standards.

For this approach to become reality requires more time because the industrywide standards have not yet fully emerged. A great deal of work has been done in recent years to establish these standards and much progress has been made. However, there are not yet robust standards in every area. Some standards are agreed upon and industrywide, but the contents are insufficient for a developer to make an efficient product using the standard. They are forced to use extensions of the standard in order to create salable products. Other standards are complete but have not yet been unanimously agreed upon across the hardware-manufacturing and software-development communities. When this full consensus is made and thorough standards are available, then all software will be developed using the standards and will be applicable across multiple platforms. It will be some years before this agreement is achieved. Many software companies are enhancing products now with standards in mind so that, once they are agreed upon, they will not have to rewrite their products to achieve the standards.

Open Systems

A great deal of confusion has arisen concerning the definition of "open systems." The term has become popular in recent years and has been taken up by a number of companies attempting to demonstrate that their products are "open." Proponents of the Unix operating system often refer to Unix as providing an open-systems approach to computing. In fact, the ideas of open systems go far beyond a single operating system or software approach. Open systems is a vision for the future where any application software — on any machine using any operating system, any user interface, and any data-base product — will be able to easily (the jargon word is "seamlessly") pass information to other systems. The key to this end is the achievement of industrywide standards as discussed earlier.

To develop these standards and have them agreed upon by all international standards bodies and hardware and software manufacturers is a complex task. Many standards bodies are government or quasi-government organizations, and there is a great deal of political negotiation between various standards bodies and private commercial organizations. Eventually, however, these standards will be developed by agreement or by necessity. The Open Systems Foundation is a consortium of standards bodies and industry leaders aimed at creating the international industrywide standards required to make truly open systems a reality. But the task is slow and laborious.

Computer-Integrated Manufacturing

Books and articles about factory automation written in the 1980s predicted that the production plant of the 1990s would be populated entirely with robots and automatic machine tools producing perfect products by the millions in a lights-out plant at very low cost. The naive enthusiasm of the techno-pundits has proven to be far more complex and difficult to achieve in practice. Factory automation is not a magic wand that solves production and quality problems. In fact, some major corporations like General Motors have poured millions of dollars into robotics and automation only to find that they did not have the management ability to profit from these new technologies. The appropriate use of automation is a powerful competitive edge — but automating errors is not an advantage.

Computer-integrated manufacturing (CIM) is another ill-defined catch phrase that has abounded in recent years. Some people give it an all-encompassing definition like "harnessing of all information required to correctly create products that comply with the business plan of the enterprise."[4] This definition is meaningless. A more pragmatic definition is the automated linking together of multiple computer technologies for the purpose of manufacturing excellence. These technologies include computer-aided design, robotics, automatic process control, automated material handling (AMH), electronic communications, computerized production planning and control, and automated warehouses.

The idealized CIM environment would include products designed using CAD systems, with the results of the designs (bills of material, production methods and routings) being downloaded to a production planning and control system for scheduling and materials procurement. Production itself would be highly automated with the manufacturing process being accomplished using robots, computer-controlled machine tools, and AMH and warehousing. With some notable exceptions, this picture is largely fictitious, although many companies have implemented elements of CIM most successfully.

CNC, Robots, and Automated Material Handling

Several types of programmable, multifunctional machines are used to perform production processes. Computer numerically controlled (CNC) machines can be programmed to perform a wide range of tasks including drilling, milling, and grinding. These machines are essentially stationary. Robots have much more movement than CNC machines and are used for such tasks as welding, painting, and moving or placing parts.

Automated-material-handling equipment is used to move materials within the plant. The simplest form of AMH is a conveyer belt with more sophisticated and complex automatic guidance vehicles (AGVs). Conveyers, used for years to move material from one workstation to another without human intervention, require a single standard route — like an assembly line — because they cannot be easily moved. Automatic guided vehicles are far more flexible because they can be programmed to move almost anything to any part of the plant. They receive instructions from a controlling computer (usually in response to an inventory-pull command) and use under-floor tracking to guide them to and from destinations. While more flexible, AGVs are more expensive and difficult to install and use.

Islands of Automation

While the grandiose vision of total automation has rarely been achieved, many companies utilize some automated activities within their plants. These innovations are often production cells employing CNC machines, robots, and AMH equipment dedicated to the manufacture of one product or series of products. These "islands of automation," while sophisticated and successful, stand alone and are not integrated with other parts of the production process.

Organizations that design or configure to order can create highly flexible production cells that make a wide range of products using the machines within the cell and programming them anew for each job. There is considerable integration between machines and equipment, and the cell may be programmed by the automatic download of instructions from a CAD system. However, this integration is contained within the cell. These cells can make products of extremely high quality and low cost even in small quantities.

Examples of less sophisticated islands of automation include automated paint-spraying cells, automatic welding units, and automated machining. Occasionally, these automated cells are linked together somewhat, creating the picturesquely named "archipelagoes of automation."

Automated Warehouses

The use of automated warehouses has become common in recent years. While the capital expenditure for an automated warehouse is high, it is frequently justified by improved accuracy, speed of material flow, space saving, and reduced inventory-carrying costs. When material is received into an automated warehouse, the controlling computer selects a storage location for the item and instructs the automated conveyers and forklifts to deliver the material to that location. The computer has location-selection logic programmed into it, and this logic can be quite complex, including such factors as space optimization, shelf-life control, and lot or serial tracking. When the material is needed on the shop floor, the computer selects which location to fill the requirement from and instructs the warehouse equipment to receive the item. The automated warehouse may also perform additional tasks like packing or palletizing the material, staging part kits, or (with the help of AGVs) moving the material to the workstation requiring the item.

Specialized uses of automated-warehousing methods include such functions as control of tool stores (which tracks tool usage and controls tooling maintenance), handling of hazardous materials, or sequencing of products. Bloomington Seating Company in Normal, Illinois, a manufacturer of automobile seats, is required to deliver seat sets in the customer's production sequence. Despite being a sophisticated just-in-time (JIT) manufacturer with no need for a finished-goods warehouse, Bloomington uses an automated warehouse as a method of sorting the manufactured products into dispatch sequence.

The carousel is another kind of commonly used automated warehouse. The purpose of a carousel is not to store items but to streamline the picking process. A carousel, a large piece of rotary equipment containing multiple stocking locations, is usually stocked with the fast-moving items that are picked most frequently. Instead of stockroom personnel traveling around the warehouse and picking items they need, the carousel automatically moves around to the appropriate stocking locations for the items required. The carousel can be automated by having picking requirements downloaded into the computer controlling the carousel. The equipment then automatically locates items in the correct sequence for stockroom personnel to do the picking, packing, and dispatch.

It is important to note that an agile manufacturer's decision to use an automated warehouse must be made carefully. A company aiming for zero inventory will have no long-term need for a warehouse — manual or automated. Some companies become beguiled by the exciting technology involved in automated warehouses. They lose sight of the fact that they are working toward the elimination of inventory and should not simply apply a high-technology Band-Aid for a traditional storage problem.

Communications

The key to CIM is communications. Even within an island of automation there is a need for machines to communicate one with another. If these machines are manufactured by different vendors, communicating effectively is often difficult because the machines do not speak the same language. They employ different protocols and instructions that have to be translated by the communications software.

Cell communications itself is a problem. Trying to link together many cells, or to provide real-time feedback from process control to production control or from the CAD system to the manufacturing-engineering system, is a much larger problem. Many companies use more than one CAD system — perhaps to handle designs for different kinds of products — and even the CAD systems do not communicate.

There are two aspects of communication networks — hardware and software. Hardware consists of the physical media used to transfer the information and includes coaxial cable, twisted wires, or optical fibers for local-area transmission. These different technologies provide different speeds of transmission, band widths, reliability, and cost. Software associated with the network will often require combining protocols and characteristics of different manufacturers' equipment.

The difficulty of communicating has been one the most significant factors preventing widespread use and acceptance of CIM technologies. There have been a number of suggested standards for CIM communications, and some standards are beginning to be widely accepted. One of the first standards was IBM's system network architecture (SNA), which is widely used to interface IBM's noteworthily different hardware technologies. Other computer companies, particularly Digital Equipment, established their own integration architecture not only to interface their own equipment but also to link it with other manufacturers' equipment. It used to be quite common to use a small DEC VAX to interface two IBM computers because DEC's VAX-SNA software was so comprehensive and reliable.

The Open Systems Interconnection (OSI) is an industrywide model rapidly gaining acceptance because almost all hardware vendors have subscribed to it — in policy if not in practice. The OSI model breaks down communications requirements into seven levels from the lowest device inter-connection up to wide-area plant-to-plant communications. A communications standard has been determined for each of the seven levels. Despite its acceptance, the OSI standard is not yet widespread on the shop floor.

General Motors took the lead with shop-floor communications in the early 1980s when the company first began a program of massive automation. Equipment being installed throughout GM's many locations came from numerous vendors each with different modes of communication. Management decided to set up a standards committee to determine standard communication protocol for use within its plants. GM had sufficient financial clout to force vendors to adhere to the GM standard that became known as the Manufacturing Automation Protocol (MAP).

The MAP standard, which began as a convenience for GM, was soon accepted as a standard throughout a wide range of industries. In 1984 a MAP User Group was formed that included a number of non-GM companies. A European MAP User Group was formed in 1985 to represent the needs of European industry as MAP standards were expanded and refined to meet broader needs of non-GM companies. MAP was formally accepted by the Corporation for Open Systems, which took responsibility for MAP protocol within the United States while ESPRIT (a research project of the European community) picked up the torch in western Europe. Now considerably extended as a protocol, MAP encompasses the seven layers set out in the OSI model and is endorsed by the International Standards Organization.

Throughout its development process, MAP has gradually become more broadly defined. MAP Version 1 primarily addressed the lower-level device-to-device communications in a shop-floor environment. MAP Version 2 specified a broadband network designed to provide multichannel, noise-resistant media that could be routed throughout a production plant. When MAP Version 2.1 was announced, hardware vendors finally came into line with the standard because it specified a consistent, stratified network structure capable of handling communication tasks from the device level to the area level. MAP Version 3 has extended the protocol further by including the communications required to link laboratory environments with production environments. In addition, better — and higher-level — programming languages have been introduced to enable the protocol to be implemented more easily within a wide range of devices and equipment.

When to Use CIM

Computer-integrated manufacturing is not the panacea claimed by some. Companies that plunge into islands of automation and then into fuller integration are asking for trouble if their automation strategy is not well thought through. Automation of all kinds should be introduced in accordance with the objectives of an agile manufacturer and only after all manual processes have been perfected. A company that automates before manual procedures have been perfected will automate wasteful activities and lose an opportunity for improvement. In addition, if any process is to be perfected, a clear and thorough understanding of the processes is required. A company that automates without detailed and practical understanding of its processes will invariably fail or (at best) be suboptimal.

Automated material handling is an area where many organizations rush in too soon. Providing a faster and cheaper way to move material is not the solution to a material-movement problem. The solution is to eliminate the need for moving material in the first place. Only after detailed waste-reduction work has been done, whereby need for material movement is brought to an absolute minimum, should automation be implemented. Rank Xerox, for example, did an excellent job of implementing agile methods into its production plant in Venray, Holland.[5] This plant reduced production costs by 48 percent in five years, improved quality from 10,000 to 750 parts per million reject rate in five years, improved delivery from 70 percent to 96 percent, and reduced inventory from three to 24 turns per year. However, the newly laid-out shop floor included a highly sophisticated automated-warehousing system and automatic guided vehicles for material movement. This state-of-the-art materials-handling system was widely admired by numerous journalists and visitors on plant tours. Ironically, only after AMH and AVGs had been in and working for several months did the Xerox people realize that the shop could be laid out in such a way as to eliminate most of the material movement and the need for holding inventories. The glamour of the technology can be a hindrance to thinking through the real issues that require resolution.

CIM should be introduced to augment the agile objectives of the organization, including the improvement of quality, better synchronization of the plant, providing additional flexibility and customer service, and/or reducing production cycle times. Robots frequently can manufacture to significantly higher levels of conformance and consistency than people, a quality improvement that makes good use of CIM. Flexibility afforded by machines that can be run briefly or for long periods, while also maintaining quality, adds to an organization's agile manufacturing capability. CNC machines or robots that can be set up very quickly through reprogramming can markedly improve production cycle times. These are the reasons to introduce CIM.

Data Collection

Most traditional companies rely on manual data entry for keeping computer systems up-to-date and accurate. The problem with manual input is its inaccuracy and the time taken to do it. Both inaccuracy (a quality problem) and the time factor (a non-value-added problem) constitute waste for an agile manufacturer and should be eliminated as soon as possible. One element in the solution of these problems is the use of automated data-collection devices.

There are many kinds of data-collection devices, including automatic counters to record completions, automatic equipment for tracking rejects or quality deviations, and automated process-control devices. The most common device for data collection is optical-character-recognition (OCR) equipment, which primarily uses barcodes. A barcode is fundamentally a machine-readable method of printing information onto labels, containers, tickets, and documentation. A number of standards now define the size and shape of barcodes for use in different situations, and manufacturers of barcoding printers and readers accommodate each standard. Indeed, some leading manufacturers were instrumental in defining the standards.

Agile companies typically use barcodes to identify the part numbers of products, components, and raw materials. Materials being received from a supplier will be barcoded with the component or raw-material part number so that any receiving or inventory transactions can be entered using the barcode, instead of having to enter the part number manually. Containers occasionally will show additional information like lot number, quantity in the container, and a quality reference code. Similarly, the part number of the finished product or subassembly has a barcoded container so that the completion quantities can be recorded easily and accurately.

There are three primary kinds of barcode readers — fixed devices, manual devices, and hand-held devices. *Fixed barcode readers* are installed at a specific place on the shop floor or in the warehouse and automatically read barcodes printed on containers or labels as the items move past the device. *Manual barcode readers* are used in retail stores where operators pass the barcoded label over a "magic eye" or use a wand or barcode gun to read a label. The guns are helpful because they can read barcodes at a distance; the operator points the gun at the barcoded label and pulls the trigger. A flash of laser light passes the information back into the barcode reading device, which converts the barcode lines and spaces into a number or a word that is transmitted immediately to the computer system. The third kind of barcode device, a *hand-held or portable reader*, is particularly useful in a warehouse because the operator can carry it around and read the barcodes. Information read by a hand-held reader is stored within the device and relayed back to the computer when the operator places the reader into a communications port periodically throughout the day. The communications port, connected to the computer, passes the information back as a batch of activities.

Many specialized data-collection devices have been designed to work under particular conditions. For example, there are devices designed to work under severe weather conditions, in dirty conditions or in a clean room, or where there is considerable vibration. Data-collection equipment often collects data in places where it is undesirable or dangerous for people to work.

Reasons to Use Data Collection

There are only two good reasons for introducing barcoding into an agile-manufacturing location: (1) for accuracy and (2) to speed up the entry of data, particularly when the information is required by the customer. The biggest cause of inaccuracy in transaction entry is the mistakes people make entering information like part and lot numbers. Of course, the software does contain many validation checks to ensure that information being entered is correct and accurate. However, many legitimate entries are still errors — they pass the validation but are still wrong. Barcoding can go a long way toward eliminating those errors because the accuracy rate of barcode readers is extremely high.

Introducing barcoding too early into the implementation of agile methods is a common and serious mistake that leads to automating transactions that should be eliminated. A recent edition of *Manufacturing Systems*, a trade magazine focused on computerized manufacturing systems, presented two case studies of companies implementing barcoding techniques.[6] One company fell into the trap of barcoding unnecessary transactions; the other, despite being a manufacturer of some hardware needed for barcoding, waited until JIT implementation had eliminated the need for the majority of transactions.

Representatives of the first company, Ciba Geigy Composite Materials, explained how they had implemented a complex MRP II system and had, as a result, created a nightmare of paperwork for labor reporting, work-in-process tracking, scrap reporting, and so forth. MIS manager Julio Castilla explained, "We were installing a very expensive and very powerful MRP II system, but it would put a heavy strain on workers to input so much data. If they weren't able or willing to do it, the whole system could fall apart of its own weight."

The solution was to go out and purchase a complex and expensive data-collection system consisting of 52 barcode readers linked through a concentrator into a PC that interfaced with the IBM AS400 running the MRP II system. Manual data entry was replaced by automated data entry, starting with time and attendance and moving through the multitude of other transactions required to feed the voracious appetite of the MRP II system. In fact, a goal of implementing the data-collection system was to not change anything. Castilla commented that one of his goals was to try to keep everything the way it was before installing the barcode equipment.

People from the second company, Norand Corporation, a manufacturer of portable computer systems, also implemented a thorough data-collection system. Barcoded inventory tickets were read by scanners connected to radio-frequency transducers, which fed the information back to Digital Equipment's VAX minicomputer running the planning and control system. Previously, Norand had implemented JIT production, including cellular manufacturing, safety-stock elimination, cycle-time reduction, inventory pull through the plant with backflushing of components, and a total-quality approach. Throughout the initial implementation of JIT, all transactions were entered manually. According to materials director Fred Seely, "Implementing JIT gradually gave us time to look at our manufacturing process. We avoided falling into the trap of automating procedures that — per good JIT practice — were soon to be obsolete." These obsolete procedures included time-and-attendance reporting, tracking of shop-floor movements, and the majority of inventory transactions — the same kinds of transactions causing such a problem at Ciba Geigy. Norand took the right approach. They (1) eliminated all unnecessary transactions before automating the processes and (2) perfected manual systems before moving to automatic systems.

Radio-Frequency Data Collection

The previous example describing Norand Corporation's implementation of barcode-data-collection devices mentioned the use of radio-frequency (RF) terminals. These data-collection devices provide considerable flexibility and power for a company requiring fast information feedback and needing the accuracy and efficiency of automated data collection. An RF terminal is a portable barcode reader with a transmitter attached. The devices are either hand held or fitted to moving equipment (like a forklift truck). Information transmitted directly from the RF terminal to the computer provides the operator with real-time feedback of validations and information. The plant or warehouse is fitted out with transducers and relay stations, usually attached to the plant ceiling, which feed information back to the "base station" or multiplexer that in turns feeds data into the central computer through some kind of hard-wired network link.

An advantage of radio-frequency data-collection devices is that they are portable while at the same time directly connected to the central computer. They do not require a batch update of information because information is updated in real time. The advantage of a real-time update is that the computer is always up-to-date. More importantly, however, all on-line validation checks and responses can be made as if information were being entered manually at the terminal. In contrast, when a hand-held data-collection device is used (without an RF connection), information is stored in the device until it is returned to its port. Then the information is transmitted back to the central computer. Validation errors or other warnings and messages required from the central computer must be reported back as a part of the batch update and corrected or adjusted after the event.

Of course, RF data-collection systems are considerably more expensive than less-sophisticated systems. Any company embarking on automated data collection must determine the value and importance of real-time updating and feedback. The majority of companies find that batch updates of inventory control or production systems are quite adequate. To ensure that all errors are captured and corrected, batch updates must be frequent and procedures must be in place to review the ensuing validation and error reports.

Integration with Manufacturing Planning and Control Systems

Integrating automated data-collection systems with manufacturing systems can be a complex task, and the essential question centers on real-time or batch updates. If real-time updates are required, then programs are needed that will emulate manual data-entry programs within the system. In fact, many barcode equipment suppliers also supply software that will drive the standard data-entry programs, tricking the computer system into thinking data is being entered manually so that the system responds exactly as it would to a manual terminal. The software reads these responses and relays them to the data-collection device. This same software also contains provisions to collect data within the controller and to update the central computer in a batch manner using the same manual terminal-emulation procedures. This batch updating would be used only in the event of a break in communications or when the computer is down for maintenance or backups.

Most suppliers of standard manufacturing computer systems prefer a batch-update approach. The real-time, on-line approach requires special programs to be written for each type of barcoding equipment, whereas a batch update merely requires that data-collection devices create interface files in a required format. The interface programs within the computer system then perform all the transaction entry-and-file updates. This approach means that interface information can be created from any kind of data-collection device, and the software company is not dependent upon a particular manufacturer of data-collection equipment.

Interface programs that read in the information from the data-collection system and update the files within the computer system can become quite complex. They must emulate all transactions that can be entered manually into the system. These possibilities will include all inventory transactions, all work-order transactions, production-completion and backflushing procedures, physical-inventory and cycle-counting results, lot-and-serial-control information, quality data, and time and attendance data. In reality, a user organization will not need all of the transactions required within a complete manufacturing control system, and the task of implementing automatic data collection is considerably reduced when the number of required transactions is reduced. Similarly, a user organization may decide to implement data collection in stages and to gradually build up the number of transactions required to run the system. This way the interface programming can be split into a number of smaller tasks instead of being all written at once.

All data-collection-equipment manufacturers have hardware and software products designed to make it easy to interface their equipment with central computer systems. Some manufacturers provide software tools that make it easy to set up the network that links the data-collection devices to the computer system. This software contains error-detection and correction features, fail-safe procedures when systems go down, and automatic backup of information, as well as features that make the network fast and efficient. Some equipment suppliers (Intermec Corporation, for example) provide the ability to download certain information from the computer system into the data-collection network. This information can be used to validate entered data prior to uploading the information into the central system. Information like part numbers, work-order numbers, and production schedules typically are downloaded into the network making the first-level validation more powerful.

Many data-collection devices themselves are programmable. Manufacturers have included an elementary programming language (often a cut-down version of Basic) into barcode readers so that data-collection equipment can be tailored to user needs. User organizations are then able to program specific validations into the device, tailor data-entry processes to include the jargon of the company and the sequence-of-information entry, and make the same device usable for many different functions. Programmable data-collection devices make automatic data collection more flexible and easier to implement and use.

Summary

While technologies and ideas of modern computing are changing rapidly, the following trends are emerging:

- Graphical user interfaces (GUI) like Windows and Motif are becoming the standard approach.
- Data is stored on relational data bases. This makes information accessible and simplifies development of application software because many housekeeping tasks can be handled transparently within the data base.
- Networks and client/server arrangements link multiple users together and make use of the computer power available on their desktops. Large and complex

tasks can be handled on the central (server) machine.
- Software must be able to be run on different kinds of computers giving users wider flexibility in choice of hardware.
- Computer systems must be open. Information must be easily passed from one system to another and reports, inquiries, and other outputs readily available to the user.

Computer-integrated manufacturing and the use of robots, numerically controlled machines, and automated-warehousing and material-handling systems are not the panacea the experts once thought. However, when objectives are in line with the goals of agile manufacturing, the appropriate use of automation can be a powerful competitive edge. The gradual introduction of communication and other standards in recent years has made CIM far less complex to achieve.

Data collection, often using barcodes, is a highly effective method of improving accuracy of transaction entries and of reducing the burden of data entry. However, care must be taken not to automate activities that instead should be eliminated.

Notes

Chapter 1

1. Orlicky, Joseph. *Material Requirements Planning*. New York: McGraw-Hill, 1975.

2. Wight, Oliver W. *Production and Inventory Management in the Computer Age*. Boston: CBI Publishing Company, 1974.

3. Wight, Oliver W. *MRP II: Unlocking America's Productivity Potential*. Boston: CBI Publishing Company, 1981.

4. Walter, William C. "The Case for an MRP System Manager." *Journal of Production and Inventory Management* 2(1990).

5. Letters. "Beyond MRP II." *Information Week*, 20 January 1992.

6. Cheveny, Robert P., and Lawrence W. Scott. "Survey of MRP Implementation." *Journal of Production and Inventory Management* 3(1989).

7. "Can America Compete?" *The Economist*, 19 January 1992.

8. Goddard, Walter E. "Toyota Versus Nissan: Two Approaches to Resource Planning and Scheduling." *APICS International Conference Proceedings*. Falls Church, Virginia: APICS, 1986.

9. Schonberger, Richard J. *World Class Manufacturing: The Lessons of Simplicity Applied*. New York: The Free Press, 1986.

Chapter 2

1. Toffler, Alvin. *Future Shock*. New York: Bantam Books, 1979.

2. Schonberger, Richard J. *World Class Manufacturing: The Lessons of Simplicity Applied*. New York: The Free Press, 1986. Maskell, Brian H. *Just-In-Time: Implementing the New Strategy*. Carol Stream, IL: Hitchcock Publishing, 1989, distributed by Productivity Press.

3. Miller, Jeffrey G., Jinichiro Nakane, and Arnoud De Meyer. "Flexibility: The Next Competitive Battle." *Report of the International Manufacturing Futures Survey*. Boston: Boston University, 1988.

4. Sohal, Amrik, and Derek Taylor. "Implementation of JIT in a Small Manufacturing Firm." *P&IM Review* 1(1992).

5. Harlock, George, and Ron Howard. "JIT at Aston Martin Lagonda: A Breakthrough Recovery Plan." *Annual Conference Proceedings of the British Production and Inventory Control Society* (1991).

Chapter 3

1. Krantz, K. Theodor. "How Velcro Got Hooked on Quality." *Harvard Business Review*. (September 1989).

2. Maskell, Brian H. *Just-In-Time: Implementing the New Strategy*. Carol Stream, IL: Hitchcock Publishing, 1989.

3. Hall, Robert W. *Attaining Manufacturing Excellence*. Homewood, IL: Dow Jones-Irwin, 1987.

4. *Ibid.*, p. 186

Chapter 4

1. Ohno, Taiichi. *Toyota Production System: Beyond Large-Scale Production*. Cambridge, MA: Productivity Press, 1988.

Chapter 5

1. Miller, Nakane, and De Meyer. "Flexibility: The Next Competitive Battle." *Report of the International Manufacturing Futures Survey.* Boston: Boston University, 1988.

2. Iacocca Institute. *21st Century Manufacturing Strategy Report.* Lehigh, PA: Lehigh University, 1992.

Chapter 6

1. Pollack, Andrew. "Doing Business by Computer." *New York Times,* 10 July 1986.

2. Iacocca Institute. *21st Century Manufacturing Strategy Report.* Lehigh, PA: Lehigh University, 1992.

3. Emmelhainz, Margaret A. *Electronic Data Interchange: A Total Management Guide.* New York: Van Nostrand Reinhold, 1990.

4. Martin, Andre. *Distribution Resource Planning.* Englewood Cliffs, NJ: Prentice-Hall, 1983.

5. Maskell, Brian H. "Order Book Review for Distribution Resource Planning." *CONTROL. BPICS* (October 1984).

6. Smith, Bernard T. *Focus Forecasting and DRP.* New York: Vantage Press, 1991.

Chapter 7

1. Schorr, John. *Purchasing in the 21st Century.* Essex Junction, VT: Oliver Wight Publications, 1992.

Chapter 8

1. For a more thorough discussion of these issues see:

Maskell, Brian H. *Performance Measurement for World Class Manufacturing.* Cambridge, MA: Productivity Press, 1991.

2. For a case study of this approach see:

Scott, Allen S. "SPC for Continuous Quality Improvement." *International Conference Proceedings.* Falls Church, VA: APICS 1989, p. 442.

Chapter 9

1. The following articles are representative of a number published in the 1980s:

Edwards, James B., and Julie A. Heard. "Is Cost Accounting the Number One Enemy of Productivity?" *Management Accounting* (June 1984).

Sephri, Mehron. "Manufacturing Accounting System for the Factory of the Future." *Proceedings 24.* Falls Church, VA: APICS, 1986.

Maskell, Brian H. "The Accounting Aspects of Just-in-Time." *Management Accounting* (UK) (September 1986).

2. Vollman, Thomas E., and Jeffrey G. Miller. "The Hidden Factory." *Harvard Business Review* (Sept/Oct 1985).

3. Johnson, H. Thomas, "Activity-based Information: A Blueprint for World Class Management Accounting." *Management Accounting* (June 1988).

Cooper, Robin. "Elements of Activity-based Costing." In *Emerging Practices in Cost Management.* Boston: Warren, Gorham & Lamont, 1992.

4. Raffish, Norman, and Peter Turney. "CAM-I Glossary of Activity-based Management."*Computer-Aided Manufacturing International* (July 1992).

5. For a detailed explanation of the ABC analysis process see:

Brimson, James A. *Activity Accounting: An Activity-based Costing Approach.* New York: John Wiley & Sons, 1992.

6. Borden, James P. "Software for Activity-based Management." *Journal of Cost Management* 5:3 (Fall 1991).

7. This definition comes from the student workbook of Productivity Inc.'s seminar, "Management Accounting for the World-Class Manufacturer."

8. Kanatsu, Takashi. *TQC for Accounting: A New Role in Companywide Improvement.* Cambridge, MA: Productivity Press, 1987.

9. For a discussion of these issues see:

Monden, Yasuhiro, and Michiharu Sakurai. *Japanese Management Accounting: A World Class Approach to Profit Management.* Cambridge, MA: Productivity Press, 1989.

Howell, Robert A., John K. Shank, Stephen R. Soucy, and Joseph Fisher. *Cost Management for Tomorrow.* Morristown, NJ: Financial Executives Research Foundation, 1992.

Chapter 10

1. Williamson, Ian. "CAD/CAM and the Business Process." Reprinted in *The Handbook of Manufacturing Automation and Integration,* edited by John Stark. Boston: Auerbach Publishers, 1989.

2. For a thorough discussion of concurrent engineering see:

Hartley, John R. *Concurrent Engineering: Shortening Lead Times, Raising Quality, and Lowering Costs.* Cambridge, MA: Productivity Press, 1992.

3. For a more detailed discussion of the use of target costing in Japanese industry see:

Monden, Yasuhiro. *Cost Management in the New Manufacturing Age.* Cambridge, MA: Productivity Press, 1992.

Sakurai, Michiharu. "Target Costing and How to Use It." In *Emerging Practices in Cost Management,* edited by Barry J. Brinker. Boston: Warren, Gorham & Lamont, 1992.

Chapter 11

1. For a fuller discussion of Unix's development see Anderla, George, and Anthony Dunning. *Computer Strategies 1990-9.* New York: John Wiley & Sons, 1987.

2. Early 1993.

3. A recent study by the consulting group Temple, Barker & Sloane concluded that experienced GUI users accomplished up to 58 percent more correct work than experienced users of character-cell systems.

4. Stark, John, ed. *Handbook of Manufacturing Automation and Integration.* Boston: Auerbach Publishers, 1989.

5. Timmer, Nico. "The Factory of the Future — Today." *Proceedings* (European Conference of BPICS, 1985).

6. Stevens, Larry. "Data Collection: An Essential Complement to MRP II." *Manufacturing Systems* (February 1992).

Parker, Kevin. "RF Data Transmission Aids JIT Implementation." *Manufacturing Systems* (February 1992).

About the Author

*B*rian H. Maskell is the president of Brian Maskell Associates Inc., a consulting group committed to assisting manufacturing and distribution companies become world-class organizations ready to meet the challenges of the 1990s. During his 20 years of experience as a manager, consultant, and software developer within manufacturing and distribution companies in England, Europe, and the United States, he has worked in assisting hundreds of companies in the implementation of advanced manufacturing and distribution systems. Prior to starting his own company, Mr. Maskell was vice president of product development and customer support with the Unitronix Corporation in Mount Laurel, New Jersey, a leading supplier of manufacturing, distribution, and financial software.

A sought-after conference speaker, Mr. Maskell has presented papers at many conferences sponsored by the American Production and Inventory Control Society, the Productivity Institute, the Profit Management Institute, and others. He has published over 50 papers and articles in journals ranging from *Manufacturing Systems*, *P&IM Review*, and

Journal of Applied Manufacturing Systems to *Management Accounting* and interviews with *Inc.* and *Fortune*. He is the author of three widely acclaimed books: *Just-in-Time: Implementing the New Strategy* (Hitchcock Publishing, 1989), *Performance Measurement for World Class Manufacturing: A Model for American Companies* (Productivity Press, 1991), and *Software and the Agile Manufacturer: Computer Systems and World Class Manufacturing* (Productivity Press, 1994).

Mr. Maskell is a Fellow of the American Production and Inventory Control Society, is certified with the Chartered Institute of Management Accountants in London, and holds an Honours Degree in Engineering from the University of Sussex, England.

Index

ALSO FROM PRODUCTIVITY PRESS

Productivity Press publishes and distributes materials on continuous improvement in productivity, quality, and the creative involvement of all employees. Many of our products are direct source materials from Japan that have been translated into English for the first time and are available exclusively from Productivity. Supplemental products and services include membership groups, conferences, seminars, in-house training and consulting, audiovisual training programs, and industrial study missions. Call toll-free 1-800-394-6868 for our free catalog.

20 Keys to Workplace Improvement
Iwao Kobayashi
The long term success of any company is based on its ability to adapt to change and remain competitive. Kobayashi, the developer of the "20 keys" system, shows companies how to increase adaptability by continuously improving 20 critical areas within the company. He describes how to recognize, measure, and strive for excellence in these areas that have a direct impact on quality, cost, and delivery.
ISBN 0-915299-61-5 / 252 pages / $39.95 / Order 20KEYS-B179

Cost Management in the New Manufacturing Age
Innovations in the Japanese Automotive Industry
Yasuhiro Monden
Up to now, no single book has explained the new cost management techniques being implemented in one of the most advanced manufacturing industries in the world. Yasuhiro Monden has taught the principles of JIT in the U.S. and now brings us firsthand insights into the future of cost management based on direct surveys, interviews, and in-depth case studies available nowhere else.
ISBN 0-915299-90-9 / 198 pages / $45.00 / Order COSTMG-B179

Cycle Time Management
The Fast Track to Time-Based Productivity Improvement
Patrick Northey and Nigel Southway
As much as 90 percent of the operational activities in a traditional plant are nonessential or pure waste. This book presents a proven methodology for eliminating this waste within 24 to 30 months by measuring productivity in terms of time instead of revenue or people. CTM is a cohesive management strategy that integrates just-in-time (JIT) production, computer integrated manufacturing (CIM), and total quality control (TQC). From this succinct, highly-focused book, you'll learn what CTM is, how to implement it, and how to manage it.
ISBN 1-56327-015-3 / 200 pages / $29.95 / Order CYCLE-B179

Handbook for Productivity Measurement and Improvement
William F. Christopher and Carl G. Thor, eds.
An unparalleled resource! In over 100 chapters, nearly 80 front-runners in the quality movement reveal the evolving theory and specific practices of world-class organizations. Spanning a wide variety of industries and business sectors, they discuss quality and productivity in manufacturing, service industries, profit centers, administration, nonprofit and government institutions, health care and education. Contributors include Robert C. Camp, Peter F. Drucker, Jay W. Forrester, Joseph M. Juran, Robert S. Kaplan, John W. Kendrick, Yasuhiro Monden, and Lester C. Thurow. Comprehensive in scope and organized for easy reference, this compendium belongs in every company and academic institution concerned with business and industrial viability.
ISBN 1-56327-007-2 / 1344 pages / $90.00 / Order HPM-B179

Hoshin Kanri
Policy Deployment for Successful TQM
Yoji Akao (ed.)
Hoshin kanri, the Japanese term for policy deployment, is an approach to strategic planning and quality improvement that has become a pillar of Total Quality Management (TQM) for a growing number of U.S. firms. This book is a compilation of examples of policy deployment that demonstrates how company vision is converted into individual responsibility. It includes practical guidelines, 150 charts and diagrams, and five case studies that illustrate the procedures of hoshin kanri. The six steps to advanced process planning are reviewed and include a five-year vision, one-year plan, deployment to departments, execution, monthly audit, and annual audit.
ISBN 0-915299-57-7 / 241 pages / $65.00 / Order HOSHIN-B179

A New American TQM
Four Practical Revolutions in Management
Shoji Shiba, Alan Graham, and David Walden
For TQM to succeed in America, you need to create an American-style "learning organization" with the full commitment and understanding of senior managers and executives. Written expressly for this audience, A New American TQM offers a comprehensive and detailed explanation of TQM and how to implement it, based on courses taught at MIT's Sloan School of Management and the Center for Quality Management, a consortium of American hi-tech companies. Full of case studies and amply illustrated, the book examines major quality tools and how they are being used by the most progressive American companies today.
ISBN 1-56327-032-3 / 606 pages / $49.95 / Order NATQM

Performance Measurement for World Class Manufacturing
A Model for American Companies
Brian H. Maskell
If your company is adopting world class manufacturing techniques, you'll need new methods of performance measurement to
control production variables. In practical terms, this book
describes the new methods of performance measurement and how
they are used in a changing environment. For manufacturing managers as well as cost accountants, it provides a theoretical foundation of these innovative methods supported by extensive practical
examples. The book specifically addresses performance measures
for delivery, process time, production flexibility, quality, and
finance.
ISBN 0-915299-99-2 / 448 pages / $55.00 / Order PERFM-B179

Measuring, Managing, and Maximizing Performance
Will Kaydos
You do not need to be an exceptionally skilled technician or inspi-
rational leader to improve your company's quality and productivity. In non-technical, jargon-free, practical terms this book details
the entire process of improving performance, from why and how
the improvement process work to what must be done to begin and
to sustain continuous improvement of performance. Special
emphasis is given to the role that performance measurement plays
in identifying problems and opportunities.
ISBN 0-915299-98-4 / 284 pages / $39.95 / Order MMMP-B179

TO ORDER: Write, phone, or fax Productivity Press, Dept. BK,
P.O. Box 13390, Portland, OR 97213, phone 1-800-394-6868, fax 1-
800-394-6286. Send check or charge to your credit card (American
Express, Visa, MasterCard accepted).

U.S. ORDERS: Add $5 shipping for first book, $2 each additional
for UPS surface delivery. We offer attractive quantity discounts for
bulk purchases of individual titles; call for more information.

INTERNATIONAL ORDERS: Write, phone, or fax for quote and
indicate shipping method desired. For international callers, telephone number is 503-235-0600 and fax number is 503-235-0909.
Prepayment in U.S. dollars must accompany your order (checks
must be drawn on U.S. banks). When quote is returned with payment, your order will be shipped promptly by the method requested.

NOTE: Price are in U.S. dollars and are subject to change without
notice.

Productivity Press, Inc.
Dept. BK, P.O. Box 13390 Portland OR, 97213
Telephone: 1-800-394-6868   Fax: 1-800-394-6286